Frederick Guttman R.

IN THE BEGINNING GOD CREATED A HOLOGRAM (THE ORIGIN)

First edition. April 15, 2024.

Copyright © 2024 Frederick Guttmann.

ISBN: 979-8231097104

Written by Frederick Guttmann.

AND IN THE BEGINNING GOD CREATED...

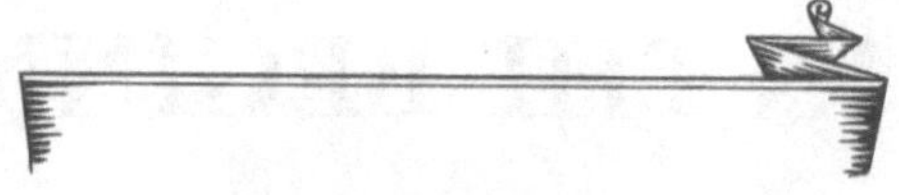

A HOLOGRAM

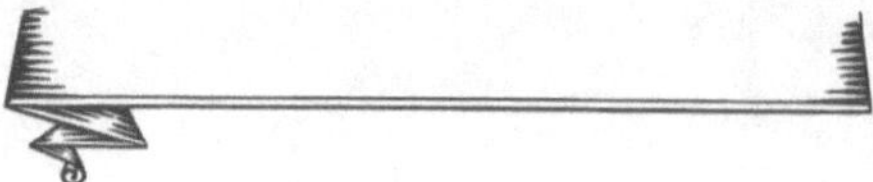

The Sakla Rebellion IV

"*The followers said to Yeshua (Jesus): 'tell us what our end will be like.' He said: 'Have you discovered the beginning to be, therefore, seeking the end?' Because* **_Where the beginning is, the end will be._** **Fortunate is he who is at the beginning: he will know the end and will not taste death.'**"
(Gospel of Thomas, saying 18)

The Sakla Rebellion IV – The Origin
First Thesis - June 2017
Second Thesis – August 2019
Cover: Frederick G.
Pages: 280
Contact: frederickguttmann@gmail.com
www.frederickguttmann.com
Candelaria, Tenerife (CP 38530) – Canary Islands (SPAIN)

Abbreviations and Particular Uses:

· *Adam:* Adam. The man.

· *Alphabet:* Hebrew alphabet.

AT: Old Testament.

· *Avatar:* body vehicle.

· *Cognate:* particle from which arises the root of a word. Several definitions can come from the same sound source or cognate, such as the sound 'Lib' in Spanish would be cognate with 'book' and 'freedom'.

· *DU:* Book 'The Disappearance of the Universe', by Gary Renard.

· *Aeon:* from the Greek Aeon, it is a sphere and state of reality where a superior deity is personified, and which also defines a time.

· *It is:* context of the spirit, or the spiritual.

· *Sphere:* context of a reality.

· *Ev:* Gospel.

· *Gnosis:* Greek word that means Knowledge, Knowledge, Understanding.

· *Heimarmene:* interstellar and dimensional region of destiny.

· *ICAR:* Roman Catholic Apostolic Church.

· *Irushalaim:* Jerusalem.

· *Jevah:* Eve, or Jivah. The woman who brings life or who is Life.

· *Monad:* uniqueness, in Greek.

· *Moshe:* Moses.

· *World:* material cosmos.

· *NH:* Nag Hammadi, an Egyptian town where in 1945 almost 50 manuscripts from the Valentinian school were found.

· *NT:* New Testament.

· *Permutation:* exchange or re-positioning, if applicable, of the situation of the letters, in Kabalah called Temura. For example, the name 'Mijael' (Miguel) by rearranging its letters can be read as 'Malajei' (angel).

· *Pleroma:* totality, in Greek.

· *RS:* The Sakla Rebellion (the previous books I have written in this saga).

· *SAO:* Service to Others.

· *SAS:* Service To Self.

· *Tanak:* Old Testament.

· *Torah:* Pentateuch (the first 5 official books of Moses).

· *Towsang:* our salt system.

· *ACIM:* book 'A Course in Miracles', by Helen Schucman.

· *Valentiniano:* from the school of Valentino (or Valentin), from the 2nd century AD. C. in Rome.

· *Yeshua:* Jesus of Nazareth.

Index

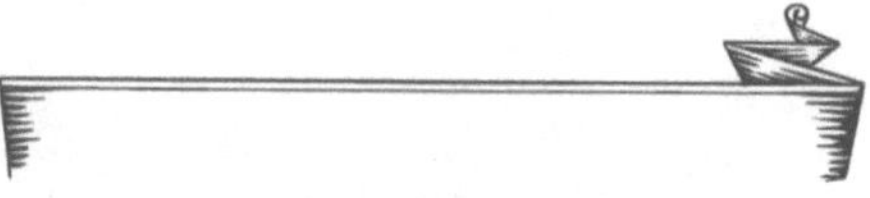

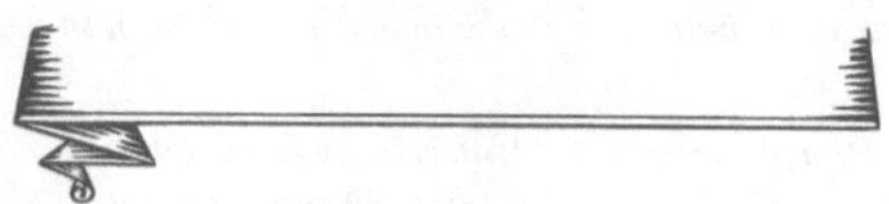

INTRODUCTION

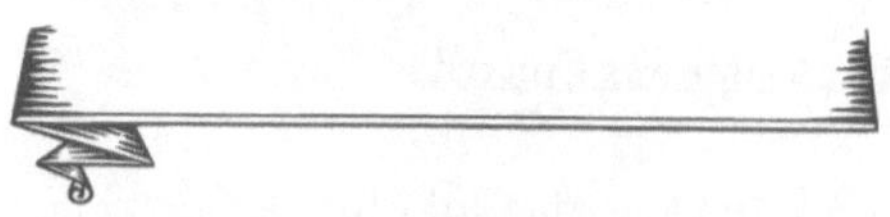

«Yeshua said: 'whoever drinks from my mouth will become like me.
I myself will become that person,
and hidden things will be revealed to him.'
(Ev. Tomás, said 108/106)

Unlike animals, we can consider that human beings give great importance to our personal reality. Certainly, other forms of life that we know on this planet do not have a sufficient level of consciousness to carry out this type of reasoning. When thinking about who we are we cannot exclude another existential doubt: 'where do we come from?'. Empirical science from the 16th to the 20th centuries would tell us that we "simply" come from certain "casual" processes in "particular" environments that resulted in various changes in forms of life, which changed psychically, morphologically and biologically, to go away. becoming completely different ones, and that thousands and millions of years later would become what we could now say we are.

For the speculative and superstitious mind, man, and the rest of existing things, were created by one or more supernatural beings from some other state of abstract reality, while others would attribute this fact to a source in other dimensions, or other planets. Mythologies are a clear example of these phenomenal and archetypal reasoning. Although, only until the consolidation of ideas about the mind that exploded with individuals such as Sigmund Freud and Carl Gustav Jung - among many others - began to understand mythology and its archetypes, and the idea of a Collective Mind. Plato and a few before him, up to Freud and Jung, had already commented on this, and had indicated the reality of Mind over the "world" of phenomena. However, the ability to fully comprehend the apparent bottomlessness of Mind was not yet entirely plausible until less than two centuries ago.

Do you wonder what the mind has to do with the origin of the cosmos? All. I am not going to address data that I discussed half a decade ago in the initial Sakla Rebellion trilogy, but I do have to remind you, if you read it, that mythology plays an important role in our understanding of the perceptible universe - as long as it is understood. their symbols and know how to adapt to the principles of the mind -. As a point of reference, I use the story that we have received by transmission from the Hebrew prophet Moses (or Mashah, if we read it literally from the Hebrew language, or Moshe, as the Jews call him). Because? Well, I am Jewish by descent, an Israeli citizen and I speak Hebrew, so what better than to give my opinion from a more scholarly angle on this field? And if you ask me, why Moses? Although, I can take some isolated source as a skeleton to build the body of this book, but what happens is that circumstances have arisen in such a way that these writings - which we could call "biblical" - have undergone fewer changes over time. of thousands of years, from which other remote sources have suffered, and this, based on the fact that few cultures as old as the Hebrews came to notify things in writing. Apart from this detail, it has been proven - especially with the understanding of the Kabalah, and theses such as the book by journalist Michael Drosnin, 'The Secret Code of the Bible' - that these texts could not have simply been invented by mortals.

Yes, the Sumerians and Egyptians wrote important things before the appearance of, for example, Abraham himself. The point is that the Hebrews did not change the written versions or the data as needed, while the common denominator in the world is that "the winner rewrites history", or a king who does not want to be a loser for future generations, changed engravings. about its history, and will leave its image and that of its dynasty, in a very good position. This is how history is made up, and it no longer lacks objectivity. Let's be honest, no town is perfect. It does not mean that because the Tanak was written by people of flesh and bones with a severe fear of deceiving - so as not to be punished by their god -, it means that everything they wrote lacked ambiguities, duality or conceptual errors. Of course they erred and made mistakes in dates, numbers, locations, names, words, ideas and, above all, in scientific knowledge, not to mention that they were extremely dual. Ergo, the essential point here is to start from something. However, Moses will not be the only one who will be the main component of this thesis. Other writings will accompany me, only without too much detail, because in addition to not having the same "purity" in the historical process of their conservation, they are complex to compare with the sources that can be preserved from them.

As you will come to understand, the manuscripts in the Nag Hammadi Library are much closer to the truth than those that were imposed and made up by ICAR. You will hear me talk about Valentinus the Gnostic, from his school in Rome in the 2nd century AD. C. He is a very important speaker in this study. One will not refute the other, since all of them will deal with their version of the dual universe. Also consider that I will use my own Hebrew words, since the translations do not have the same value as in their original language. An example is that I will use definitions such as: Torah (or "doctrine", which is generically used in that context to refer to the first books of the Bible, which are attributed to Moses, and are known as the Pentateuch), Tanak (the 3 sets of texts that form the so-called Old Testament), Barashit (Hebrew name of the first book of the Bible, known in the West as Genesis). Now, the Genesis account (a Greek word meaning 'generations') is supported by dozens of other sources around the world. Moses himself in Jubilees, as well as Enoch, Baruch or Ezra, address the same points, do not mention some others, or add characteristics that had not been addressed in the others. In general, all together give us the sum of information for this thesis.

At the beginning of Genesis, the Hebrew word Barashit (beginning, created) appears as the first word, and is followed by: 'Bará' (created), 'Elohim' (god, gods, strong ones), 'et ha Shamaim' (the heavens) 've et ha Aretz' (and the Earth). The additional article 'Et' did not need to be added. Why did they put it in then? 'Et' is a restatement; It is to reliably ensure that it refers to "that" specifically. But, above all, this 'Et' is an accusative case particle, and therefore, adds the direct object. The "accusative" is not the case of the cause but of the effect. What does this mean? That the created Heavens and Earth were not the "root", but the "consequence". The Elohim creates those Heavens and that Earth as a result of something that the story has apparently not included. But as you will see by reading this work, what is not exposed – at least at first glance in the narrative – as the reason for said creation, was and is, in reality, the source - and that source is neither in time nor in space -. Just think about it for a moment: if there is no space, how can there be time? If there is no time, how can we talk about space? Imagine yourself in "nothing", in "emptiness", and you want to calculate "time". How do you do it? Suppose you create an object in that void. You could already define that there is a "time" between your apparent location and the object, which it takes to reach a certain "speed". Still, we would have to have perceptual measures to measure time. Who defines a second or a minute, or an hour, or a year or a century?

How do I come to all these conclusions, what sources do I base them on? Through various ancient and contemporary literary sources, as well as through Kabbalistic deductions. The Kabalah is the use of 5 methods to discover hidden mysteries, both in the Tanak and in the phenomena of the world and in dreams, and that is applied to the Mind itself. I have added this essay to the comparative study in light of the famous book 'A Course in Miracles' and similar works. Thus, it is easy to understand the dynamics expressed by Yeshua in the book he transmitted to Helen Schucman between 1956 and 1965, which has been published and shared until today by the 'Foundation For Inner Peace' (or 'Foundation for Inner Peace '). This work (ACIM) was carried out by the psychologist Helen Schucman channeling Yeshua, and being helped in the work of writing and subsequent teaching by Bill Thetford, Robert Skutch and Ken Wapnick. Said work (UCDM) could not be previously shared by Yeshua due to more than 15 centuries of obscurantism and Catholic repression, due to the high level of global illiteracy and because certain parameters of psychology had not yet been explained (this knot was undone thanks to works such as those carried out by Carl Jung - such as the famous book 'Archetypes and the Collective Unconscious' - and Sigmund Freud, among many others).

After this advance and the discovery of the Nag Hammadi manuscripts (in December 1945), which were not destroyed by the ICAR (Roman Catholic Apostolic Church), a large part of the true message that Yeshua wanted to transmit and that was not possible was known. understand due to the radical traditions of the Jewish community that determined the birth of so-called Christianity. The texts from Nag Hammadi (Egypt) were hidden by Saint Pachomius upon learning that Theodosius had decreed at the First Council of Constantinople (382 AD) the prohibition of the teachings and texts where Yeshua spoke about the preexistence of the soul, reincarnation, transcendence and enlightenment and spiritual freedom. Since the First Council of Nicaea, by order of Constantine the Great, a first "Bible" (canon) was established, later translated by

Jerome of Estridon into Latin, called the Vulgate. But this version did not include most of the fundamental references about the transcendence and illumination of the soul or about revelations given to the patriarchs, prophets and apostles of the Hebrew people, due to the difficulty that it would mean for the ICAR to be able to keep the people under control.

In addition to this, they removed 20 books from the official canon, 25 quotes from the gospels that were allowed to be incorporated in which Yeshua spoke of sensitive topics for the ICAR, they progressively added words and references that did not originally exist in the Greek manuscript, and they took advantage of the differences between Hebrew, Aramaic, Greek and Latin to enter translated words at your convenience. In addition, they decreed excommunication to anyone who studied these texts, such as the disciples of Valentinus (from the Gnostic school of Rome in the 2nd century AD, which until then had been the most versed in the understanding and study of the mysteries taught by Yeshua to his apostles), and any mention of these matters as "heresy". Once the Course (ACIM) was published - in 1965 - its dissemination began, but the growth and spread of information took a while to expand, especially due to the explanatory complexity of the Course. For this reason, Yeshua sent two of his apostles, Thaddeus and Thomas (presented as Arten and Pursah) to transmit to Gary Renard the mechanisms of the Course in an understandable way, which were taught from a first visit in Maine (USA). in 1993. From then on, through a dozen appearances, they gave him the content for a first book: The Disappearance of the Universe. This was published in 2002, and was followed by two more works that complete the trilogy: 'Your Immortal Reality' (2006) and 'Love has not Forgotten Nobody' (2013).

Arten and Pursah explained to Gary that the awakening thought system taught by the course (ACIM) - which is the same as that of the Buddha (Buddhism) - was not accepted by the Jews when Yeshua wanted to teach it - although some wanted to understand it (but they did not do it, just like their brother James, the Apostle Peter or Paul of Tarsus), and Philip and Mary Magdalene understood them perfectly (Thomas, Thaddeus, Andrés and Esteban were the ones who later almost understood the Message) -, but its combination with Jewish mysticism explains the entire structure of manifest and non-manifest reality, and accelerates the path of awakening and the application of mental exercises that modify reality. This was confirmed by Arten and Pursah as well as the work of psychic Jane Roberts, who was channeling Seth throughout the 1970s, and from whose recordings she published 8 books ('Speak Seth'). These works were also supported - and form a whole - by the works of researchers Carla Rueckert, Don Elkins and James Allen McCarty of L/L Research, who published 'The Ra Material' in 1982 (5 books that complement the revelation of the Oahspe, of 1882, by John Ballou Newbrough), among other groups that have been receiving updated messages from consciences from other worlds and other dimensions (as in their time other peoples received them, especially the prophets of Israel, or the famous Edgar Cayce).

Kabalah is a Hebrew word whose meaning is 'reception', that is to say "what is received", referring to the Tanak (the 3 sets of books that make up the so-called Pentateuch, in Greek). The Kabalah has 2 mechanisms: one pragmatic and the other symbolic. The pragmatic is divided into 3: Gematria (numerical meaning of Hebrew letters and words), Temura (anagrams and permutations of the letters of a word) and Notaricon (reconfiguration of new words through the union of initials or final letters of a word). Biblical quote). The symbolic method has 2 areas: Bereshit (study of the book of Genesis and its analysis from the understanding of the symbols used in the text) and Merkaba (analysis of celestial or non-physical phenomena and their archetypal meaning). This book complements the courses

I teach on Personal Growth (Transformational Development) and Kabala (UCDM-K).

Part I.

Part I.

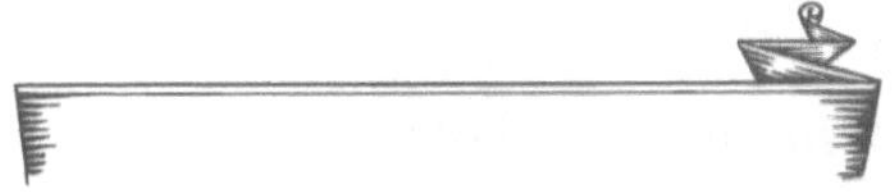

BEFORE THE BEGINNING

«The Father made his abode in the [Son] and the Son in the
Father:
This is [the] kingdom of heaven.
(Ev. Felipe, verse 96)

DAY MINUS ONE (-1)

As a civilization we have had an evident interest in knowing our origin and the reason for our existence. How can we go back to the past to know the ins and outs of what arose when there was no human way to record what happened for later generations? Do you believe in revelations? Do you think that other more advanced civilizations from another world or another dimension, or from the future, could have been giving us knowledge? In itself, over the last two centuries, scientific advances and theories have increased considerably. Although many estimates are based on assumptions, mathematical calculations or probabilities, it is accepted that models that present the idea of the origin of the universe have, to a large extent, a high percentage of possibility of being right, and coincide with what the myths of the ancients represented.. However, conventional science and much of the Abrahamic religious community supposedly do not agree in their views, and this seems to create a huge gap in ideology or popular belief. The vast majority of Christian and Muslim groups – and even Jews – reject the official

18

version of classical science, arguing that the origin of the universe is already presented in the biblical book of Genesis, in its first chapter, and does not correspond to the theory that It is supposedly attributed to scientists, who are pigeonholed as atheists.

On the contrary, in Hinduism this discrepancy does not seem to exist, even in the Catholic leadership, considering that the study of both currents has simply been misinterpreted, and they do not have to be incompatible or mutually exclusive. Many refute that a Hebrew (Moses) could have had the resources or knowledge to know the origin of the cosmos and of life to capture it, for example, in the so-called book of Genesis - much less any other human being even a hundred years ago -. In contrast, whether supporters of creationism or panspermia (especially defenders of ufology) are - to a greater or lesser degree - convinced that an "alien" intervention (angels, gods, extraterrestrials, interdimensional beings or time travelers) provided to the peoples of the past (not only to the Hebrews) of a knowledge that only now is science beginning to discover. If so, could the first chapter of the first book of the Hebrew Moses really deal with a fully "scientific" worldview? And I'm not saying evolutionary science, but pro-panspermia science (theory defended by the illustrious Fred Hoyle that assumes that life was brought to Earth, it did not occur here).

בְּרֵאשִׁית בָּרָא אֱלֹהִים אֵת הַשָּׁמַיִם וְאֵת הָאָרֶץ׃

With these Hebrew characters the biblical creation story begins, where transcribed it would say, " barashit bará elohim et ha.shamaim ve-et ha-aretz ", but what do all these words mean? Barashit, or Beresheet is the Hebrew name of the book of Genesis, and comes from the same root as the Persian 'Bahashit' (heaven). This word means "beginning", in the form 'ba-rashit' (head), where 'Rosh' is head, beginning or leadership. The word 'Bará' means 'to create' or 'to form'. That is, 'c reó'. This word is formed from the letters abyad Beit, Reish and Aleph, where the root can denote an associated hidden meaning of Bor (Beit, Vav and Reish), which is "well" (compare many other cosmogonic texts). Precisely the first word is made up of the same initial letters as the second: Bará = Bara-shit, thus being an evocation of a principle of creation or establishment (because 'Shit' would come from the voice 'Set' (establish)).

For its part, the term 'elohim' encompasses any concept related to a powerful being, a divinity or a deity, whether singular or plural. Then, the word 'Shamaim' refers to "heavens", starting from the Semitic words it is understood as a compound idea: Sham-maim (there-waters) or Shem-Iam (name of the sea). The ancient Egyptians considered the night sky as a sea, and this was a very popular idea, so it was accepted that this celestial "sea" was full of life. While sometimes words like 'Shamei' or 'Shamai' (heaven) are used, the truth is that almost all the time the plural 'Shamaim' is used, implying that beyond the celestial vault there are many "heavens." Thanks to the large number of texts that exist, we can understand that these heavens are all types of spaces, whether physical or immaterial, and of various dimensions and universes, not limited at all to the common idea of many monotheistic aspects. In the case of the Sumerian version of the creation story, the origins tablet says, "E-nu-ma e-liš la na-bu-ú ša-ma-mu," referring to Nabu as the sky, and the Sam.ma.mu as the fact that it had not yet been established.

But what would happen if we considered a variation in the initial letters and their meaning, which coincided with the Persian font? If Barashit were a deformation of the Persian Bahashit, that is, of the idea of Heaven, we could dare to formulate an alternative translation that said: "Heaven created gods, heavens and earth." Most mythologies usually begin their cosmovisions by arguing that the "heaven" created everything, and that heaven was the first of the gods, as would be the case of the Roman Caelum, or the Greek Ouranon. That means that there would first be celestial gods who would have created the celestial regions, according to the remote legends of the peoples of the Earth. Another formulation of the letter game could be based on an ancient way of writing that consisted of not separating the "words". Thus we could read from the first portion of the initial sentence, for example, with the Aramaic form: "bar ash itbara alahim atah", which means, "[the] son [of] fire created [a] you gods". Moreover, we could read: "bara shit bara elah iam atah", which would be, "created foundation, created deity, you are sea". Are you sea? The word Yam, like sea or ocean, encompasses the idea of the material universe, or deep space, the basis of matter, or multiplicity. In fact, Yam is an abbreviation of Yom, which is both day and period of time, era or cycle.

Suppose that a primeval and original Heaven existed before the time and space that we think we know, and from it the Elohim was created, as well as later "heavens" and "earths." Let us take the word Fire as an idea of infinite light. Ergo, we would have as a result that everything came from the light, who had a Son who created the Elohim. The light would be first, then his Son and then the Elohim. The other interpretation is that of "someone" who creates a foundation or base, a "place", "space" or "stage", also creates the deity, which, if not read as "elah iam atah", but as "elahim atah", would be "you are Elohim". According to any of these versions, Elohim is not the first to appear, but a later result of a source. Personally, I wouldn't consider it self-centered if the deity itself were introduced to begin with, since it doesn't tell us where it came from. Going back to the Sumerian version, in it, the word 'Sa-Ma-Mu' (whose Hebrew equivalent would be 'Sha-Ma-iM') does not seem to be translated as Heaven, but as "named". In that way? The word Shamaim is made up of the structures 'Sham' and 'Maim', where the root of Sham is the Semitic Shem (name), from the Akkadian Shemu, from the Sumerian Mu. 'Sham' also means "there", and 'Shama 'is "over there".

There are many more curiosities to delve into here, and one of them is the fact that the form 'Et' – or 'At' – is the formation of the first and last letter of the Hebrew alphabet, which in Greek would be Alpha -Omega, or "beginning and end". It could thus explain another level of understanding about the sky derived from primordial emanations, as well as the earth derived from primordial emanations. If Bahashit is the pristine heaven of Ethe, the subsequent Heaven and Earth are the upper world and the lower world that followed. In effect, that next Shamaim would be the image of the Bahashit, and the next Aretz, in turn, the image of that Shamaim. Both a later Shamaim and a material Aretz will pass away, as the Bible says on a couple of occasions: "heaven and earth will pass away." Therein lies the hidden meaning of the letters 'Alef' and 'Tau' that form the aggregate 'Et'. We could read it, then, like this, alternatively: the primordial sky created the beginning and end of the next sky, and the beginning and end of the world.

The Sky and the Water

But if we look at verse 9 of chapter 1 of the Enuma-Elish (Sumerian version of creation), it refers to heaven as Sa-Ma-Mi. In verse 8 he defines 'name' as 'Su-Ma'. These word games are key, especially if it is understood that these languages could be put together anagrammatically. Su-Ma agrees with the Hebrew 'Shma' (hear, listen), in the sense of naming something or calling someone. For this reason, even in modern Hebrew, a name is said "Shem", and it is applied from the beginning as the idea of someone's nickname, as well as their "renown" (re-name) or identity. In Sumerian it was also the idea of someone's destiny, and they understood it as the destiny of someone who came from heaven, or where they were going. The Akkadian word, then, would have come from the Sumerian Sha-Ma-Mu (named sky), as a single idea, not two separate things. What is a "named heaven"? A sky defined, established, identified, fixed and determined. The opening part, 'E-Nu-Ma' (when what [is]) and E-Lis (what is high), gives us another clue. The form 'E-Lis' derived to Akkadian as 'Elih', and this to Hebrew, where its idea of "strong" was, as among the Canaanites, the structural idea of a "god". From there, theoretically, the plural sound 'Elohim' was born. For this reason there are analogous forms such as the Arabic phoneme 'Alah' (Allah), which is "elevated" or "high" in the Hebrew language.

I assume that you have clearly heard of the famous word 'Manna', but you may not have known that it means "what is it?" (Hebrew 'Man'), or the name Moses, whose root is the cognate 'Mah'. That 'M' is a symbol of water, like Moses who was pulled out of the waters, or the Manna that fell from heaven. Exactly, the so-called Logos-Word of God became "flesh", in the sense of food to nourish, and because he defined himself as "the manna that came down from heaven", this is because 'Man' is the sound root of many ideas of "man," as in English, Egyptian, or Hindu. Now, the text speaks to us of these "heavens" or "waters there." Sky and water? Yes. In Hebrew water is a plural, "Waters" (Maim), with the letters Mem-Yud-Mem, meaning "upstream", "heaven between" and "downstream". In this universe the projection is based on water, and outer space is an example of water. Without going any further, when Dark Matter was discovered, what until now was a speculative idea became real trying to understand the shape of a galaxy that was seen from Earth as curved in an incoherent way. When analyzing that this curvature could be an undulating recline of the light, they asked themselves, what did the light bend? The explanation came from the same effect that we see when looking through a glass of water: the image behind is distorted.

Dark matter is, in effect, energy from matter that occupies space, although with the naked eye – and with current measurement equipment – it cannot be measured yet. This energy that matter that we do not see has is similar, in analogy, to the difference that we would have between light air, dense water and very dense solid matter. The air, in this analogy, would be the Dark Energy, or Ether. The Mem (or 'M') is therefore the symbol of **Matter.** The Holy Spirit has been revealing it in many ways. That is why the Hindus call her

' **M** aia', in Gnosticism they called her ' **Mary** ', and she is defined in science as 'M'. Do you remember Einstein's formula? E=MC2, the expression of the Theory of Relativity. That is, energy is equal to mass times the speed of light squared. In other words, mass is energy, yes, but more correctly mass is energy in the sense that it is energy as part of a photonic whole. In other words, matter/mass is photonic energy in concentration, and this is by focusing a consciousness outside of our perception. Atoms are photonic structures (of light) in concentration combined with quarks and other particles such as muons or bozones, and they, through this "concentration" or focusing, attach the particles that form the molecules, of which matter is made up.

The "speed of light" is an expression that was greatly popularized by the Star Wars saga, but in reality it is not a unit of measurement but of time, corresponding to more than 1,000 million kilometers per hour (specifically 1,080,000,000). We do not say that something is 5 million light years away in the sense of distance but in the sense of the time that the light would take to reach that place from us. Considering that light has mass, and constitutes mass, we must equally conceive of the fact that time and light are linked in the perception of the reality of the universe. Earth time is subject to the rotational and orbital movements of our sphere, so the perception of continuity and the cycles of time that we want to associate with the universe would only be measured according to our units of measurement. For this reason, the concepts and measures that are seen, for example, in the Bible, are not understood, since the interaction with other "times" or measures and cycles that belong to the "heavenly dwellings" is not understood. This is the case of 1 celestial hour, which according to a certain Heimarmene calculation corresponds to 41.66 Earth years, and which adjusted below the equivalent measurement of 60 minutes could be close to 40, which is a widely used unit of time. in ancient times (see RS3, the Temporal Aeon - page 68).

A consciousness produces vibration, vibration produces frequencies and waves, and these produce energy, and the energy, in an intelligent and organized state, is focused to produce the photon, and therefore the most basic of electromagnetic fields: The atom. The atom then perpetuates this same pattern through consciousness and forms molecules, such as hydrogen or oxygen, from which water and other chemical elements are subsequently derived. Water is the state of matter that starts from the condensation of gases where its atoms reduce their speed of movement (just as the solid is to an almost immovable point). However, the Hebrew expression that the sky is 'Shamaim' ascribes that "up there it is water." The Hebrew letter 'Shin' is a symbol of fire and energy. By writing 'SH-MYM' we are looking at a quasi-mathematical formula of energy (Shin) per water (Mem-Yud-Mem), or mass. Water is made up of oxygen and hydrogen molecules. The Greek word 'Idros' (Ydros, form "ydros-genos") means "water". If we consider MYM as the formula for water, hydrogen would be 2 particles of Ydros with one of oxygen, which is the main component of the atmosphere. The atmosphere expresses the elemental idea of heaven.

So, MYM are two principles of the idea of water-hydrogen, and a principle of heaven-oxygen, which is H2O. Clouds of interstellar gas were formed by hydrogen, from which matter condensed to form planetary bodies, and from which the elements of this hydrogen fused, giving rise to helium, which is the basis for the formation and existence of stars. As a result, we have that Shamaim symbolizes the opposite state of matter within this universe, and that it has the potential essence of energy, more than what can be seen in water with the naked eye. However, the difference between Shamaim and Maim is that Shin is the intelligent energy source of the Maim. In this way we can conclude here that Intelligent Energy, Ether or Dark

Energy - whatever we want to call it - is the conscious-energetic base that establishes matter in the cosmos. And since consciousness is vibration, therefore, the material world is a replica of the superior world, where consciousness is absolute, making clear the absence of consciousness of the being in itself, or lacking self-conscious state.

Everything is Energy

Consequently, there is no vacuum, there is no nothing in space in the sense of nothing as nothing. There is consciousness as intelligent energy even if a saturation of objects is not seen. If we understand this mysticism of the Hebrew letter Shin, it leads us to its number (300), the same as the combination of the letters Reish (200) and Kuf (100), which form the definition 'ReK' (empty). Ergo, the void is not a nothing in the sense of an absence of anything, but merely the non-existence of a condensation of elements that give rise to a perceptible material structure for us. In other words, the fact that there is no material form visible or measurable to us does not mean that there is not something. There is energy in everything, energy in everything, because everything is energy and is maintained by energy, but not electrical as we perceive it, but intelligent, as consciousness (or 'with science'). This is the same if we analyze the brain. As an organ it has its tissues, but within it there are electrical charges, or synapses, where the neurons do not touch each other. The planets and stars do not "touch," but they are actually connected. The brain is a structure, but its cells do not perceive it as something outside. For both the cells of the whole body and the cells of the brain, the body is a whole and the brain is a whole, and each of them is a part of the whole. They do not have an idea of separation, but rather they think as a single network, as a single structure. When the cells stop seeing themselves as a whole, the body gets sick. For example, a case of this is the manifestation of cancer.

Just as our entire body is energy, the entire universe is energy. But understand it from the greater perspective: I am not speaking on a poetic level, but rather a literal one. This is exactly as applicable to the fact that the entire universe is a body. Planets, stars, galaxies, beings... all of them are the atoms, molecules and cells of said body. They are all connected like atoms are in a structure. Each atom is each atom, but it is part of a structure or electromagnetic field that forms a body, be it a cell, a plant, an insect, a man, a rat or a dinosaur. The atom is the photonic condensation (light) of the consciousness of this universe through vibration, acting as a single organ. But it happens that the great brain is not the body, but only the part that processes neuronal, nervous impulses. That means that in this analogy the universe is just a portion of a larger structure, and that larger structure is a consciousness or spiritual body called 'Elohim'. In the treatise called 'The Discourse on the Eighth and the Ninth' (Nag Hammadi) Hermes Trismegistus says that "the universe received a soul", how so? Identity or personification, self-awareness. However, it existed first, but then it began to "live", or in other words, it became individualized. He had the "permission" to be something separate and to become self-conscious.

I am not the first to recognize that there are paradigms that had not previously been questioned about these words, more than the Genesis story itself, but there is something that has been less speculated on: if Moses tells us how everything arose, what is it like? that does not tell us either about the origin of water or about the identity, source and nature of the Creator? Where did it come from? If we want to search for this enigma, we cannot start from the sources used until now. Let's see, we have observed that this creation that seems to be mentioned in these texts does not seem to reflect a beginning but a subsequent appearance of something based on a certain thing that occurred previously, or derived from a certain place or previous state. Traditional mythologies are not useful to address this matter. All of them are subject to post-bigbang approaches, not "pre-bigbang", or could it be another way? These people could have been taught by aliens or time travelers, but would those aliens or time travelers know what was before the Big Bang? I call the initial state "Day -1" or "Day minus One". From this 'Day -1' state comes the 'Day 0' state, which is the state before "being named" as 'Day One'.

The Parody of Christianity

Now I must emphasize the development of this theme, given its relevance. For this I will cite Nag Hammadi, where the ICAR did not arrive, and could not destroy the knowledge that from 382 d. C. stopped accompanying us to learn about this and many other mysteries. 70 years ago, a series of 45 manuscripts reappeared near the community of Nag Hammadi, Egypt, which were much closer to what Yesuha and his direct followers had transmitted. In December 1945, what the Valentinian school hid from the arbitrariness of councils such as the First of Constantinople, and from the orders of Theodosius, was known, translated and quickly popularized. According to my opinion, as passionate about ancient texts and scrolls, and agreeing with the criteria of serious and professional sources, the Nag Hammadi manuscripts are the Holy Grail of Yeshua's teachings that managed to survive. Yes, sorry to be a killjoy, but the history of Christianity is a Jewish fabrication based on the criteria of a man who did not personally know Yeshua (was not a disciple), a macho fisherman who was scapegoated by Rome, and a Jewish leader. religious who did not believe in his brother. That is, I mean Saul of Tarsus, Simon Peter and James, son of Joseph the architect (who tradition confused with "carpenter").

These three men, pillars of Christianity, were the ones who least understood the Buddhist nature of Yeshua's teachings, and as a result they created Christianity as an extension of the eschatological ideas of Judaism, since they saw in Yeshua the Messiah who would establish the militant Jewish dualism that he had fallen with Roboam (son of King Solomon) and Jeroboam. If you want to know the truth, don't stick to any religion, because even Christianity is a dual vision of God, and it doesn't agree with the original teachings of Yeshua in a high percentage. I say this very much to your regret - if you are a Christian - because Yeshua was not a Christian (not in the ordinary sense of the term). Although he was a Jew, his teachings were pure non-dual, that is, they were of Buddhist current, but even more accurate than those of the Buddha himself. It is rather for this reason that they did not understand him, and still do not understand him. The years that elapsed from the crucifixion to the configuration of the biblical canon were accompanied by many fables that the devotees added to the ecclesiastical narratives. There were no books (that appeared from the era of the printing press). Thomas was the one who began to take note of Yeshua's teachings. Of the 4 synoptic gospels, 3 come from a source called 'Gospel Q', which in his time was known as 'Words of the Master'. When that book disappeared, only one complementary source remained, which was the gospel of 'Los Dichos', later known as the 'Gospel of Thomas'.

Of those 3 synoptic gospels, only Mark's was in person, but that Mark was merely a child when Yeshua preached, and he was such a young disciple that, apart from the years that passed until this was written, he would not remember too many concepts, which Besides, I didn't understand. The synoptic gospels were written after Paul's letters, therefore, they are reminiscent of the Pauline doctrine, rather than the doctrine of Yeshua. Yeshua's teachings were not understood by the Jews, not only because they did not want to accept it, but because they simply did not understand it. The ones who really understood her at the time were Mary Magdalene and Philip, as well as 4 other followers, the 7 being great friends of adventures, laughter and teachings who even traveled the world together before Yeshua began his so-called ministry. I make this introduction so that you get an idea of how different the story is compared to what we have been told, and how much the original story differs from what was later taught by Christianity. Yeshua and Mary Magdalene were in Paris, in Stonehenge, in India, in Alexandria, in Tibet, in Greece, and they were followed by Philip, Andrew, Stephen, Thomas and Thaddeus.

These were the ones who best understood the Master's words. For this reason, Philip's gospel is the most powerful gospel of all that survived, as it testifies to what Philip himself heard and understood firsthand. Next is the gospel of Thomas, which was not completed because its writer, the apostle Thomas, was murdered. What was known about Yeshua was through oral transmission, since Thomas was one of the few who knew how to read and write, and he took his time putting the Master's words into writing. As the years passed, there were many urban tales that were added to the story of events. Thus, the common gospels became a story based on real events, not on real events themselves. Stories were created like Yeshua had turned a couple of loaves and fishes into thousands of them, when in reality it was a parable about trusting in providence; Yeshua was said to have cursed a fig tree or made a whip to drive merchants out of the temple, an aspect of the character they wanted to capture to give it a touch of boldness. Even things that John the Baptist said were confused with Yeshua, such as "love your enemies."

Above all, words here and words there passed the years and decades and what was remembered was mixed with the beliefs of Judaism about the 'Messiah Liberator of Israel', clearly through the prism of a hero with a divine touch who in any moment would free the special people from the oppression of their enemies. When Judas son of Cariot (Ihudah Ish-kariot) saw that Yeshua was not showing what he expected, he devised a plan with supporters who wanted to overthrow the Jewish aristocratic structure and raise a new political order based on the Zealot vision, which, with Yeshua's help would destroy the Roman Empire. Yeshua knew that this needed to happen, even though it was not his intention, and he knew that a drama, like that of a crucified god, would be more transformative through the centuries, than that of a second Buddha.

The plan of Judas's supporters worked, and to this day the conspiracy was believed. Yeshua showed himself to his followers 3 days later to explain what had happened but they did not understand him - and they did not want to either - because his mind only imagined that the Messiah had returned in flesh and blood from the very underworld, to legitimize his sovereign power. More and more dualism and strengthening of the dream of illusion. However, Mary Magdalene and Philip did understand it. María Magdalena, Felipe, Andrés, Esteban, Tomás and Tadeo continued their friendship and worked on the work 'Words of the Master', however, this writing was a heresy for ICAR. They could not accept preaching or enlightened women, like Mary, who showed the same abilities as Yeshua. Thomas and Thaddeus had seen them walking together on the water when they were not even 17 years old (Yeshua), and 15 (Mary), as Thomas himself explains in Gary Renard's work, 'The Lives in which Jesus and Buddha Met.'. After Yeshua, Mary was the singing voice, and she continued to be, but given the machismo of her society, and Simon Kefa's (Peter) animosity towards her, she followed his teachings in Paris. However, the name Paris derives from 'Par-Isis', where Isis was the Egyptian name for Mary in the ideological sense (it has nothing to do with the Isis wife of Osiris who were masters of Atlantis 12,000 years ago, and also for several cycles were lords of the Earth).

The third of the enlightened at that time was Philip - who knew Yeshua better than all the others -. Both he (Philip) and the disciples of Yeshua and those closest to him knew the mysteries that he had transmitted. For almost 11 years after the so-called "crucifixion", Yeshua returned to teach them about the mysteries of the eons. They received constant revelations of things that conventional Christianity did not get to know. The ICAR feared that this truth would be known by the vulgar, and in this way they would lose power over the masses. Consequently, they took in 382 AD. C. what had been written about Yeshua and the sect of the Nazarenes, and made up. The ICAR eliminated all the quotes where Yeshua explained the cycles of reincarnation and our ethe origin, and established the dogma of perpetual hell, and whoever did not accept the infallibility of the pope and his dome (the curia) would be excommunicated. These arbitrariness often reached the extremes of physical torture and death. As the writings were studied and revised, words were substituted, others removed, and more and more added, especially to enhance Yeshua's non-human character, so as to legitimize the new deity vision of the Christian religion. Thus, appreciations such as that the Son is the manifest, the Father the real, and the Holy Spirit the abstract, were distorted, leaving aside the vision of uniqueness, and replacing it with a confused, disordered and incorrect doctrine about three beings. divine in one

Now, what greater background does Christian revisionism have? All. The so-called Christianity – or let's correctly call it 'Pauline religion' – only legitimized the Jewish religion – or let's correctly call it 'Israelite nationalist theocratic religion' – by putting the focus of reality on the material world and elitism, promoting more ideas of better than some others, glorified matter or the "struggle" and "death" of some "bad" people. Apart from this, the real teaching of Yeshua was relegated to the merely poetic and philosophical, considering more the opinions of Paul of Tarsus as a structural foundation. Faith was subordinated to a strictly religious belief system about the existence of a single god; but above all the dual vision of that God was strengthened by considering him promoter of the murder of his own son. The Christians completely ignored that Yeshua did not really need to let himself be killed, and that he did not pretend to teach anything about death. In a way, Christian doctrine understood that Yesuha's action had the function of implying that death is not real, and it was partly so, but it was not the foundation of the symbology that Yesuha wanted to convey.

Judas Ishkariot and others planned a conspiracy to save Yeshua's life, since they knew that a large number of members of the Sanhedrin wanted his death. The plan was to persuade one of the supporters of Judas' vision to fully believe that he was the Messiah. There were at least two lines of space-time continuity from when that Passover meal took place until Yeshua physically projected himself to the apostles and made them believe that he had marks of the crucifixion. Yeshua was so advanced on a psychic level that he created a collective hologram where everyone present believed they saw him die, when in reality the one who died was a substitute. Yeshua simply, upon saying goodbye to them at dinner, became his Fourth Density self (the most advanced state of Resurrection possible in a Third Density state), and from there he accompanied from Fourth Density a man whom the followers of Judas They drugged Ishcariot and convinced him that he was the Messiah and had to go through a torture to fulfill the prophecies. Yeshua was unable to explain the truth to his disciples - who were shocked to see him and believed he was a ghost -. He ate so they wouldn't think he was a ghost but since they didn't understand he left the subject at that.

He already knew that they would not understand him and that his thought system would only fit the circumstances as a "resurrection" of his physical body from the dead. Only Philip and Mary Magdalene understood at that moment that he had passed into Fourth Density and was showing them a projection of his self as if he were still Third Density. Why hasn't this been said before? Who was going to accept it? The eastern world? They knew little or nothing about Yeshua. The western one? Until just a few decades ago, almost everything was under the inquisitive and inflexible power of the Catholic Church. And even if it became known, how is it explained? Who do you explain the Truth to if it transcends the perceptible world? How do you explain to him about mental holograms, collective mind projections, space-time continuity alterations? It was not something done with technology. Yeshua created a parallel reality with mental power, using the collective consciousness. They saw a man crucified. Yeshua made it seem like it was him because he could introduce an archetype, which is a whole system of codes based on the idea of the Cross.

The man they drugged was Jewish, but he did not survive the crucifixion, and his followers stole his body to say that the Messiah had revived. Yeshua had nothing to do with the conspiracy, nor was any of that stuff about sacrifices part of his teaching system. Sacrifices, suffering, death... That was contrary to the love, peace, oneness of God and happiness that he was teaching as ultimate truth. He knew what Judas was planning with others, and for this reason he told him, "what you are going to do, do it now." Yeshua had transmitted everything when he had to transmit and that up to that moment it was possible for it to be transmitted. Judas and those who were with him wanted to make the Sanhedrin guilty in the eyes of all the people so that the people would ask for the dissolution of the Sanhedrin. The idea was to create a change of Jewish government based on Zealot politics taking advantage of the pretext that the Sanhedrin had killed the Messiah, and then they would say that he had revived and ascended to heaven. They wanted to make seem fulfilled the prophecies of a Liberator who would motivate them all to overthrow the Roman tyranny and occupation. Yeshua's message was not to die for anyone. He never said that. That was an interpretation of Paul of Tarsus in light of his previous beliefs. Yeshua wanted to use the symbol of surrender to teach that truth is not dual. The message of the cross is that if you are attacked (crucified) you don't respond. You don't repay evil for evil, you turn the other cheek. Love is surrender, it doesn't stop evil or fight, because it would make it real. Only love is Real.

Judas had the idea to save Yeshua, to preserve his life, believing that by making the Sanhedrin then Yeshua would have the free way to confidently address the people as a leader. Yeshua knew that this thought system of the ego was going to materialize at any moment - because the ego is predictable -, and he saw it in his mind beforehand and knew that this drama could be used positively to transmit a greater message of love through the "non-fight" (not facing your enemy is recognizing that there is no enemy). Yeshua had seen the ideas of the Zealots for years and was well aware of the beliefs of the Essenes, and was aware of how Judas and others expected him to be the one to arm the Jews and use his powers to destroy Rome, whom the Jews Essenes identified with Belial. I knew that at any moment these people were going to plot something to get their way and it was only a matter of time. Out of love, he escorted from 4th Density a man whom Judas and his fellow conspirators persuaded was the Messiah, and that man believed it.

Such was the drugged state that he could not even bear the cross and it fell off, to the point that a certain Simon of Cyrene had to help him. That man began to regret the idea he accepted when the effects of the drug wore off when he was hung. He believed what they told him: that he was the Messiah, and that his action would be a milestone. He thought that God would save him at any moment, but seeing that it did not happen, he said, "Father, why have you abandoned me?" Yeshua used what he already knew was an obvious act of the ego that would happen sooner or later (he already saw it coming, and if it wasn't Judas it would be someone else, because the ego reacts with tragedy and suffering when it is attacked). I knew that the idea would promote the so-called belief system of Christianity, establishing certain guidelines that could be of some use, even if they were not pure non-duals. He knew that this would be of some use until a generation more evolved in understanding Mind came along. If the synoptic gospels are properly analyzed,

Yeshua never actually said that he was going to die. What is recorded in these texts is that he would have said that "the son of man" would go through this and that.

Who was with Yeshua in Gethsemane to witness what supposedly happened if those who were "watching" were not awake, but asleep? Yeshua was alone, according to what the story says, but if Yeshua did not tell anyone about this and there were no witnesses, where did they get that this happened? How did they know that Judas had sold Yeshua for 30 pieces of silver or that he would have repented and returned the pieces? If Judas attempted suicide and fell down a ravine, at what point did he confess? What most do not question is how it is assumed that things could be reported where there were no witnesses. At that time they did not have cameras. You could confuse one person with another, and Yeshua himself was apparently confused with Thomas (for that reason they called him "the Twin"). The point is that the surrendered person looked like Yeshua and Yeshua himself was making things look like what they wanted to see. It's not something easy to explain. If you do not understand what the Collective Mind, Collective Consciousness, the mental configuration of reality, and other complex issues of mentalism are, it is difficult to understand what happened.

Furthermore, who defines that a prophecy was being fulfilled or had not already been fulfilled, or that the prophecy was actually being properly interpreted? Arten and Pursah explain that what was announced in Isaiah 53 referred to a character who appeared a few years after Isaiah. The temple that Yeshua claimed would be destroyed is the true temple, and the three days are three cycles. In reality, apart from what the gospel of Thomas tells – and perhaps general ideas of the gospel of John – nothing told in the synoptic gospels can be taken literally. There were too many cultural additions, urban stories, and when all of this was converted into "gospel" – in the sense of biography – those who had been there had already died. The real point, what is the important part, is that Yeshua knew perfectly well that the world is a dream, that nothing but God is real. He laughed and enjoyed himself because he was fully aware that nothing in the world (the cosmos) is real. If someone thinks about suffering, pain, loss or death, he is part of the delusion. No film director loses his reason while they are shooting their movie believing in the role that the actors are playing. Suffering does not exist for those who have freed themselves from deception. They want to understand things that require time. If you do not begin to indoctrinate your mind in the full and absolute conviction that the world is not real - instead of seeing it in a philosophical sense - you will continue to make judgments, and to make judgments is to be deceived, since you believe in something unreal.

Therein lies the main confusion with the meaning of the crucifixion, the 3 days of death and the resurrection, that is, the myth that Yeshua revived his Third Density state. If you tell someone that he will be resurrected, you are lying to him, because neither flesh nor blood can inherit the kingdom of God. The resurrection is of the Mind, since it is the Mind that has died. The body was never alive, because it is not real. The resurrection of the dead refers to humanity, since it died when testing the tree of Torah, that of good and evil, which is the tree of judgment, the one that created this universe. The theme of resurrection simply lies in the fact that the being awakens its state of the next density, and that is a mental change. If you are in a state of Third Density and you activate your state of Fourth Density, you only have to mentally program the way in which you will get rid of the body that until now covered you. If you have a Third Density body, for example, you simply mentally define the way in which you will leave it to activate your Fourth Density body. There are Fourth Density beings in this generation, in Third Density bodies, as there were before Lemuria, more than 35,000 years ago.

This is due to the transition that the Mayans had announced, which elevated our planet to Fourth Density. Thus, in this age of Aquarius, humanity begins to see the collective elevation to Fourth Density, in line with the Earth. The few years that a being that has already activated their Fourth Density state, but still belongs in a Third Density body is a period of mere peace and analysis, like when you already know that you have received an inheritance and it is only a matter of a few years for you enjoy it. You reach the end of the judgments and have forgiven everything, then you are no longer afraid and you enjoy the complete peace of God. It is illogical to assume that a Third Density body revives as if it could thereby access the "kingdom of God" in the ether. What consistency does it have to say that Yeshua revives his Third Density body and then is elevated to heaven to go with the Father, if the Father is Spirit? So why revive that body? Christians assume that hope is reviving matter, something completely different from what Yeshua taught.

Yeshua's message was not based on his followers ending up "believing", well, what are we talking about believing? For them, believing in the Christ (Messiah) meant recognizing that he was an earthly king who would inherit the throne of David and gather the dispersed of his people and strengthen his nation, re-impose the law, free his people from their enemies and have an enduring kingdom over other nations. This is hard and pure dualism. A nice ego strategy. Except for Felipe, María, Andrés, Esteban, Tadeo and Tomás, that was what the rest expected. On the other hand, Yeshua and these six friends of his did know that the Christ represents a Universal Consciousness, a kingdom that is not dual, however, it is not of this world. Felipe did understand that the dispersed ones are all the beings of the cosmos that were introduced into the projection after the Big Bang. Mary of Magdala and Philip did understand that the Messiah would free them from the illusion of the Dream, not from its images (enemies). Yeshua ate with his followers after the drama of the crucifixion simply so that they would not believe that he was a ghost. He told them, "I am not dead," but they did not understand. How was I supposed to explain the truth to them?

Peter said "I don't know who that man is", why? Not out of fear, but because I didn't recognize that this was Yeshua. In Gethsemane Judas himself had to give the sign of a kiss so that Yeshua would be recognized. Did he have a better view than the members of the Sanhedrin and the soldiers? It is absurd that Yeshua's followers were with him for years, knowing his face, his voice, his way of walking, his clothes, his gestures, etc., and then, during an entire hours-long trip through Emmaus, they did not recognize him walking his side, nor later eating together. And Maria? The Madeleine arrives first before anyone else before her husband and, standing before him, does not recognize him, but instead asks him, "Where is the master, have you taken him?" Some did not recognize that man because they knew he was not Yeshua, and that is why they did not go with that man to Calvary. Yeshua was no longer of this material, and therefore they no longer recognized him the other times he appeared, including the first appearance he made when they were hiding and materialized out of nowhere saying "shalom." He disappeared at the Emmaus meal before the eyes of those present and did the same on that occasion. After the last supper Yeshua finished his work and ignored what was going to happen next, to the extent that he left it stated by stating, "look, this is my flesh and this is my blood", not what they were going to see the next day. following.

Yeshua was not a sidekick with Judas, he simply 'took advantage' of the conspiracy to incorporate an impressive message that would mark a relevant historical fact for generations, until the day the Catholic Church allowed freedom of worship and psychology opened the ground for the understanding of mind. He does not violate Free Will. That was going to happen anyway, and other than that, he already knew that almost no one understood or would understand his teachings and his background, so the Holy Spirit, who knows the script of everything, joined him in using a simulation. who presented an education system whose purpose was non-fighting. Thus, in the mid-60's Yeshua began to transmit 'A Course in Miracles' (A Course in Miracles, or ACIM), to the psychologist Helen Schucman, in a 9-year process. When one reads it, the Holy Spirit confirms that it was transmitted by Yeshua, and that it represents the revelation of the Truth, only possible thanks to a process of historical changes that have allowed it. The Course recalls that the Mind created the cosmos, which explains the things that Yeshua said at the time 2,000 years ago, but with a "Shakespearean" touch of a psychological nature.

Day Minus Two (-2)

Based on what has now been exposed, let us consider that it is plausible that there are other more reliable sources of knowledge – as far as the Truth is concerned – to arrive at that TRUTH. I recommend as the main source of Truth the revelation of Yeshua given by Helen Schucman, as A Course in Miracles, and after it the revelations given by the apostles Thomas and Thaddeus to Gary Renard in his works The Disappearance of the Universe. After them I usually recommend the Seth Material, the Ra Material, the Oahspe, the Nag Hammadi manuscripts and other works of the Valentinian school, 'The Secret of the Flower of Life' by Drunvalo Melchizedek and the works of Hermes Trismegisto or Thoth (as The Emerald Tablets). After this one can, then, understand the reality that surrounds us, who we are, where we come from and where we are going, and we can perfectly complement it with the works of Carl Jung, Sigmund Freud, quantum physics or the literature of the fathers of ' Foundation for Inner Peace'. It is thanks to these works that I can provide you with this thesis, and through which I will continue to summarize for you the mysteries of the kingdom of God and the essence of the universes - especially this one in which we are -.

I addressed the dynamics of primal chaos in 'The Sakla Rebellion I', but now I delve into what came before. There are, in that sense, a couple of texts that reveal to us what was before the Big Bang, and these expose that there is only one Invisible, which is Everything. This totality that Is, receives the Greek name of Pleroma, that is, the Plenitude that others have called 'Plentum'. He emanates lights that are portions of his being, and all together they are one, a unity or uniqueness called in Greek 'Monad'. The first projection of himself is known as 'Barbelo', which could be a word with Aramaic nuances, according to the form 'Bar' (son/daughter) 'Belá' (without, absent), or as 'Bal' (lordship). This could represent that he is an emanation not considered a son, or one who projects his power. The two of them produced a son whom they called Khristos, and he, asking them for their light, created an offspring whom he called Esefech, and later a prototype of man, called Adamas, who was the father of the first human of the first race of that kingdom., known as Set.

Those were the Imperishable Kingdoms that were created from those first emanations. All of this occurred before this universe, and outside of time and space, in a celestial-etheric state devoid of forms and images. Then we began to talk about "son", which in modern Greek is called 'Gios'. The Gios is, however, the source of the Giin that was produced later, which is also known as Gios or Gaos, or Jaos, from the Sanskrit form from which the word 'Chaos' comes, as well as the Greek 'Gia' (Earth)., Gaia or Gea, from the root 'Gi'. This 'Gi' symbolizes the matter, the result. First is the invisible or spiritual, then the visible, or material. The 'Gios' is the producer of the 'Gi' although he later interpreted humanity through his myths which was the other way around, with the Gios (son) being derived from Gi (the earth). For this reason it was said that man was formed from the Earth, when in reality it was first the earth (matter) that came from the son. Let's not make the mistake of assuming that I am talking about Jesus Christ. This book is not a Christian thesis. Yeshua (Jesus) is an unequivocal symbol of what I speak of, but not the source.

In the universe there are symbols, since these are representations of the apparent levels. In this way we are taught through symbols-ideas, and we can understand the lower things with the higher ones, and the higher things with the lower ones - because by seeing some we understand the image of the others -. For this reason, Yeshua is part of a series of symbols that aim to explain things on other levels, not idealize, worship or honor the symbols. Because they are just symbols! The first Gios/Yios is part of the Pleroma and the Monad, of the Plentum and Unity, of the Plenitude and Oneness. Then other Gios (sons) come, as replicated ideas of a unique reality. Like cells. The cells are replicas of others, but they are not outside the whole, the body. We can understand the propagations of the initial emanations by understanding the subsequent emanations, or vice versa. The Invisible One, who is Perfect, and the One and Complete Spirit, is fully aware of himself. He projects his reality onto his own reality.

I will use geometry to explain the mechanism: Suppose that He, being completely total, would be represented as a perfect circle. How does it become a circle? Well because he is aware of himself. If you are aware of what is in front of Him, behind, below, above, to your left and to your right, you have 6 directions, which are all, where you are aware, and that awareness is absolute, then it is all-encompassing. A consciousness that occupies the 6 points is, in itself, a sphere. The 6 points are interconnected creating a holographic network where the energy of thought is condensed. This produces a sphere. The ONE projects his own consciousness in front of Him, doubling the sphere at the limit of his consciousness, that is, in front of him, since he actually has no limit, since he is complete and absolute.

This duplication occurs at that edge, and from there emanates the center of the next sphere. The configuration of these two spheres is called 'Vesica Pescis' or 'Vesica Piscis'. This is the 'First Action'. A Second Action takes place, where another is replicated from the center of both spheres, which is the first Gios (Son), called Hristós, the Anointed One, since he is anointed and impregnated with the uniqueness of the ONE and Barbeló. Now there is a trio that symbolizes the Monad. That is, the Pleroma takes the form of the 'Tripod of Life'. It is "of life" because everything begins with three.

This totality is known in the Kabalah as 'Ein', which most closely translated would be "there is no", "without", "absent", or the principle of nothingness-silence that is everything, and whose number is 711 equal. than 'Adon' (Lord). From it, according to the perspective of Jewish mysticism, 10 spheres (sefirot) or nodes are derived that would project the abstract idea of the Pleroma as archetypes of reality. This representation is called 'Etz Jaiím', or Tree of Lives, where the 10 nodes are interconnected in 22 paths, which would correspond to the letters of the alphabet. The "Tree" is divided into 3 columns, or pillars: left, central and right; and in 3 levels: upper, intermediate and lower, which some see as 4 levels, according to the states of reality: Ein, Ethe, Atmos and Corpor. This is not a literal system but a representative one, to exemplify encompassing aspects of existence, God and the human psyche. The idea of Kabalah is useful to illustrate the archetypal science of concepts, beginning because in Hasidist (ultra-Orthodox Jewish) mysticism everything came from the 'Ein' (Nothing) that is fully conscious, intelligent and empowering. We can relate this idea to the Perfect One of Gnosticism. From 'Ein' comes 'Ein Sof', like the Perfect in infinity; Then comes from Ein Sof the 'Ein Sof Aur' (or 'Ein Sof Or'), which is the infinite or illimitable light. This is the Pleroma of the great primal Monad and by which everything exists, and in which everything is, which is the Primordial Father, the true trinity that is one and the same, where Perfect, Barbelo and Christ are complete.

Barbeló is the first "power", the Previous Thought of everything that came from the mind of the ONE Father, the Protennoia. She is the 'Dója', which some transcribe as 'Doxa', the Glory, or supreme state of magnificence. She asked the ONE for virtues, and one after another she received them. Since she is the Glory, and personifies the Previous Thought, she is already the first of the great virtues, and the following were added: Proigoúmeni Gnósi (Previous Knowledge), Athanasía (Immortality), Aiónia Zoí (Eternal Life) and Alítheia (Truth).. In this way the image of the Great Invisible Spirit was projected. These 5 became the First Humanity, not in the sense that humanity is now understood, but in what Adamah really means. This totality, of the 5 with the One in itself, are a 'Dekada', or totality of 10, called 'Páteras', or in the Hebrew language 'Aba' (in the Aramaic 'Ab'), which means in Spanish 'Father'. '. I repeat, they are all copies of each other. The higher projections are the root of the lower projections. The projection of all things is fractal, it is repeated at all levels.

At each level there is Adam (man, humanity), there are gods, there are fathers, there is Christ, etc. It is a projection, like a movie. Remember that the film is not a real or solid object, it is a projection of something that has already happened within time (it has already been recorded, edited and presented), which is not occupying space (because they are pixels of three colors combined in such a way that the television shapes it as moving images and in more colors) and that it did not come out of the TV (the device only shapes an emission of waves that come from elsewhere). It could be said that illusions propagate through the ether as a replica of an original source, just as radio and television waves propagate through the atmosphere. After her comes the Gios, or Yios, who was a ray of light from the Perfect One entering Barbeló. Ray of Light in Greek is 'Aktína Fotós', that is, yes, the Father took a photo of him, and that photo was the Son. Do you want more examples of analogies? There are no coincidences. The Holy Spirit is the only thing there is when there is no ego. Therefore, all truth comes from the Holy Spirit, and he encodes everything in numbers, letters, words, shapes, colors, languages, concepts, symbols. A photo is a possible embodiment by Fos (light), a projection of light.

It is important to consider that in ancient manuscripts the idea of 'Uios' as Son appears more: Without the letter Sigma - the single root 'UIO' - represents the Son. The son is the representation of the trinity or triad that is One in Totality. That is why three letters are key here. The Greek letter Ypsion (or 'Upsilon') "uppercase" is confused with the "lowercase" Gamma, which explains the association of 'Gio' with 'Uio'. The 'Ypsilon' identifies the High, the 'Iota' symbolizes the universe and the sky, and the 'Omicron' is the eye that sees everything, that is, it is aware of everything and makes reality exist. In this way, the three, or 'UIO', are one, and the Son represents the three. Remember, this does not mean that IHVH (Yahweh, Jehovah, Iao, Iaheveh) was created there. We speak of the sphere of emanations, before this current cosmos and its gods. Understanding this explains, among other things, the analogy of symbolism that the consciousness of the deity

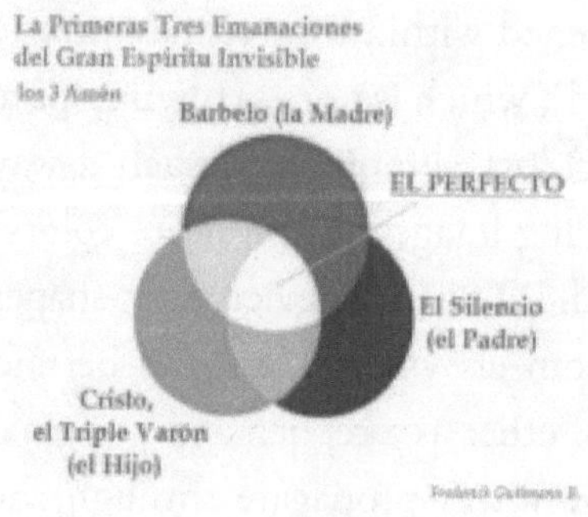

Iaheveh and the master Yeshua had.

We then have that after the Vesica Piscis another sphere emanates, which is by anointing ('jríka' in Greek, or 'mishjut' in Hebrew), producing a Third Action, beginning by receiving Nous, or 'Mind'. All this occurred in complete Silence, since everything comes from Silence, and through Silence one returns to the source. The Son asked for 'Logos' (Word), and with it came the 'Tha' or 'Thélima', that is, Will. Now this Christ, or Anointed One, was proclaimed among these virtues and emanations as 'Theo', that is, God. This offspring was able to understand the true nature and identity of that spiritual universe, and was God produced by himself in that reality or aeon. Once that reality was projected, and the Third Eonic Action was developed with all its manifestations, the Fourth Action was developed, which were already 4 integrated spheres. I have already spoken about the details of these emanations and their kingdoms in videoconferences and in the 'RS' saga, so my purpose with this chapter is to summarize and integrate the essential aspects that shape the Barashit. It was not possible to understand how the Barashit was produced, nor why, if the root of what took place prior to its appearance is not understood. Although, these details can also be studied in 'The First Treaty of Set' and in 'The Secret Book of John', both of the sources available and translated from the Nag Hammadi Library. Then, after the 4 integrated spheres comes the fifth, and then the sixth, which becomes the first great structure from which the rest are derived... Egg of Life, Fruit of Life, Flower of Life.

These glories are emanated from Barbelo, which is the Doxomedon, and the state of reality of the Divinities, called 'Kalyptos'. There dwells the self-begotten Son, and his son Esefech, and the glory of the virgin and thrice-masculine man (Christ), called Youel, or Yoel. It is from this Kalyptos that the kingdoms of the aeon of the Son arose. And this is where the aeons of Kalyptos exist, which are 4 sources. The First Aeon in it, whose first light is, is called Solmis and is with the Revealing God; the Second Aeon is Akremon, the ineffable, together with the second light, Zachthos and Yachtos; the Third Aeon is Ambrosios, the virgin, together with the third light, Setheus and Antiphantes; the Fourth Aeon is the one who blesses the great race, and with whom is the fourth light, Selda and Elenos. This kingdom is identified with the 'Adam Kadmon' of the Kabalah, that is, the Pigeradamas, the Primordial Adam. This is the first and highest of the 5 levels of reality of the Kabbalist worldview, but in reality it should be associated with the state Atzilut, which is that of emanations. The way to see it in Hasidism is Adam Kadmon as the soul state Yechidah, whose basis is consciousness; then there is Atzilut, where the soul state would be Jaiah, and the base would be the Mind; Next would be Briah, whose soul state would be Neshimah, and its base would be Thought; then there would be Yetzirah, with the soul state Ruach, and the basis of Speech; Finally Asiah would be found, where the soul state would be Nefesh, and the base would be Action. In reality Adam Kadmon would be in Atzilut, or state of emanations, and as a result of the principle of Thought made active. However, I will not go into philosophizing about the differences or coincidences between the various points of view on this.

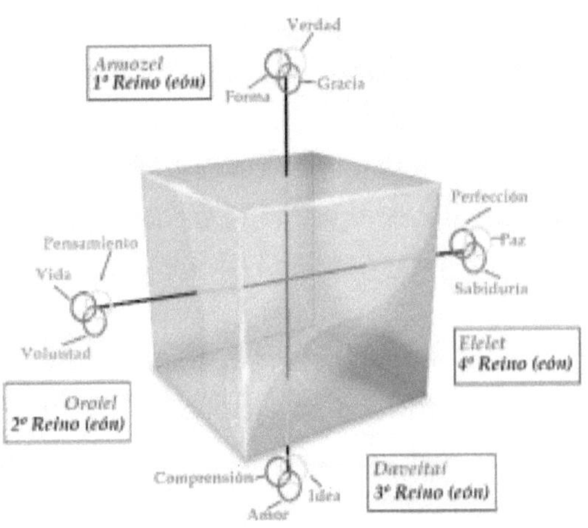

According to the Nag Hammadi manuscripts, the Yios/Uios produced a race, and received kingdoms from the upper eon invisible to him, with them proceeded angel-princes, angel-kings, faculties, powers and assistants. These beings are generically called 'Asterei', or in Hebrew 'Kojab', which in Spanish would be "stars". In short, 4 main kingdoms, each with 3 sub-kingdoms, and each kingdom and sub-kingdom with their respective powers. In the first group of the 4 came the proto-man, called Adamas, which means "indestructible" in Greek. His kingdom was called Adamantine, which also means "made of diamond." In some parts of the Gnostic texts he is called Geradama, that is, 'who produced the Adamics', or Pigeradama, who received an unconquerable mind from the Invisible (hence his name).

In the second group of the 4 kingdoms lived Set – the son named by Pigeradamas -, whose name means "foundation", "place" or "foundation". A pure spring source is said to exist in a kingdom here known as Shdom, which is in Amorrah. As a symbol of the fall, these names were given to two cities in the Zoan Valley, north of Judeah, and were transcribed as Sodom and Gomorrah, respectively. Now, in the third of the 4 groups, the descendants of Set were established, the Setites, those known as the Kodashim (saints). Their souls were destined for the third kingdom. But, in the last group things change. Regarding those kingdoms, the first star is called Harmozel, which is the first group of the kingdoms; the second group of kingdoms is from the star Oroiel; the third star is called Daveithai; and the fourth is called Eleleth.

Due to the appearance of an "angel" under the title of Elelet, in a literary work attributed to Norea - daughter of Jevah (Eve) - one can combine the idea that certain higher consciousnesses do not have anthropomorphic bodies, but are manifestations of light. These 4 come from the Mother, not the Protophanes (Adamas), as they are a thought from the perfect mind of light. In this way, the stars – like everything else in the cosmos – have consciousness, but in their case, specifically, one of the highest states of vibration. In the 'Keys of Solomon' this association of angels-princes with stars is also appreciated. We are especially interested in the last kingdom mentioned, that of the star Eleleth (or 'Elelez'), where there are three kingdoms called Perfection, Peace and Wisdom (who in Greek is known as 'Sophias').

In this way, before the stem of light there are 4 stars and their 12 kingdoms. The analogy of this with the 12 constellations that surround us and the division of our annual time into 4 stages, dependent on the steps of the sun across the face of these 12 stellar figures, is evident. In this way, our universe is a reflection-image of the universe of the primordial Christ, or god produced by himself, and from where comes the structure of the 12 saviors from whom the 12 apostles take their being and origin (see Ev. Valentino, chap. 31). This projection is manifested in the lower levels through the symbol of the planet Uranus, which is also called Heaven in Greco-Roman mythology. A year – or cycle – consists of 84 Earth years, which is an equation of 12 periods of 7 years; it is the same as saying that a cycle of 7 terrestrial is a cycle of the 12 of Uranus (see RS3, p. 173).

What happens, then, to the inhabitants of the kingdoms of Eleleth? These kingdoms were destined for souls who were ignorant of divine Plenitude. It seems strange, but that's how it is. The 'Apocryphon of John' or 'Secret Book of John' gives us to understand, through the mediation of Yeshua, that the inhabitants of these kingdoms of the star Eleleth would be temporarily ignoring "divine fullness." What does that mean? That a state was established for a ¼ part of this "social consciousness" that was understood to not be aware "at the beginning" of the truth, that is, of the Plenitude. We must understand that the understanding of Fullness is essential for discernment regarding the Totality of which we are part. The absence of this scholarship is an essential factor in dualism, an idea of separation. If we are not aware that we are part of a Whole, we will feel separated, and that brings mythological death, that is, the various types of death. By 'death' I am not referring to the ambiguous idea of abandoning a body.

Death means SEPARATION in the reality of the absolute. Leaving a body is not death but transition. Death states are, for example, suffering, illness, ailments, misfortunes, abandonment and attachment, etc. Here comes in the fact that verse 16 of chapter 5 of the 'Apocryphon of John' tells us that those destined for these kingdoms of Eleleth would not have repented immediately, "but continued ignoring for a time, and then repented later." This is so even though they are also creatures that glorify the Great Invisible Spirit.

It seems strange that Father One would be involved in a dual idea. However, it is not in the way the ego wants us to perceive it. Yeshua explained this in a metaphor, saying: "God's Divine Law is like a person who had good seeds. His rival came during the night and sowed weeds among the good seeds. The person did not let the workers pull up the weeds, but instead told them, 'no, because you could pull up the weeds and the wheat along with them.' For on the day of harvest, the weeds will be visible, and they will be separated and burned." (Ev. Thomas, 57) As described in the synoptic gospels, Yeshua explains this allegory in the light of their understanding, but religion sees in "the evil one" an idea external to us that intervenes because God allows it. Because? They say that to test us, even when Jakobo refers that "when anyone is tempted, he should not say that he is tempted by God; because God cannot be tempted by evil, nor does he tempt anyone; but each one is tempted when he is drawn and seduced by his own lust." (James 1:13-14, NT – The Bible)

The ego shields itself by looking outward, towards demons, towards the supposed defects of others, or it justifies things by saying that it is a "test from God", instead of the mind taking responsibility for its own thoughts. These "weeds" are the ideas of separation, of the ego, of feeling different from others, well, if we are all one, what coherence does it have for "I" to think about being different or alien to others? Precisely the ego is what religions personify in the idea of a devil. The archetype takes shape in your beliefs as if it were something outside of you and not a part of our consciousness. Others say that "it possesses us," when in reality it is nothing more than our way of thinking not aligned with the Holy Spirit. Look at the example in the Egyptian language, where the sound 'Sata' meant "generated serpent", "generative serpent" or "son of the earth". Everything is the same, always. The "earthly" or "worldly" going is precisely the thought form of separation, the non-spiritual or the Right Mind. The form 'Sat' is to use, and 'An' is 'earth' or 'evil', being – once again and as usual – an allusion to the thought state of separation, the perception of dualism.

We do not have a war against something outside, but primarily against something inside. That war and darkness of our being is what is projected outward, and what as Mene Collective we project as darkness is manifested and personified in genocides, epidemics, wars, natural disasters, invasions, etc. Things on a collective level are projected towards global generalities, and our internal thoughts are projected outwards into particular realities. As little Sofia (Pistis Sofias) said later, becoming aware of her actions: "Protect me, light, because bad thoughts have entered me." (Ev. Valentino 8:17) The state of guilt that exists in our unconscious promotes the need for sacrifices, while Yeshua does not ask for any of this, as he himself refers in ACIM: "In one way or another, every relationship that the ego engages is based on the idea that by sacrificing himself he becomes great. The "sacrifice," which he considers a purification, is in fact the root of his bitter resentment." (Chap. VII – 6, 1) Therefore, when the apostle said that "we do not wrestle against flesh and blood but against principalities, powers and kosmokrator of the regions of aera", he was not only speaking of the level of form and the level atmos, but from the level of Mind (since those " **ark** - <u>o</u> n- **t** - <u>e</u> - **s** " (archon) are our own psychological " **ark** - <u>e</u> - **t** -ip <u>o</u> - **s** " (archetype)).

The 5 Places of the Monad

Now, I must specify that there are 4 realities: EIN, ETHE, ATMOS and CORPOR. Ethe is the potential and source of everything, which we can derive in 'Ethereal' or 'Eternal', the upper part of the 'Is' state (the 'Spirit' which is a veil that divides the upper realities Ethe from the lower ones and its sublevels) ; The sublevels exist through the Psiky, Psijí, or Psyche, which is the potential of the Mind and the soul forms (souls, personifications or individualizations of consciousness), which the Oahspe calls 'Atmosphere', but I abbreviate it as 'Atmos'., so that it is not confused with what is known in the West as atmosphere; Finally, there is the heaviest and densest state, which is that of forms and images, and which is an atomic and molecular coating of the electromagnetic fields of the psyche, called Corpor, since it is associated with the "corporeal" idea, or of the bodies, which is the matter, or the physical. The root of the All, which is the absolute ONE, is the source of all Ethe, is the one who emanated the First Place, which is defined as the 'House of the Father', 'the Pledge of the Son' and the 'Power of the Mother'. This is the First Father of All. They are all Silence and have no beginning. From them derives the Second Place, which is that of the First true Demiurge. He is the First Overseer, the source known as 'All Father'. He provided the first ennead, that is, the first set of the 9: Gnósis, Life, Hope, Rest, Love, Resurrection, Faith, Rebirth and the Seal. These two places are in the Interior Pleroma, that is, inside the interior totalities.

From the Second Place the Outer Pleroma, or outer worlds of Ethe, is projected and surrounds 12 depths, or Dodeka. These 12 depths are 'The Image of the Father', or 'Mirror of Everything'. It is about the all-knowing depths from which all sources have come; the all-mystery source of mysteries; the all-gnosis source of all gnosis; the silence in which all silence is; the insubstantial door, source of all insubstantial; the ancestor of all ancestors; an all-father and self more removed from all paternity; the one who is the first invisible source of all the invisibles; and the truth from which all truth has come. These are the 12 depths surrounded by the Second Place, that of the First Demiurge. Behind this there is a Third Place, which is that of the Second true Demiurge, who is called 'Father', and from whose breath everything that came into existence was nourished. This place has 4 doors, each with a monad (oneness), and each with 12 dodékadas (units of 12), 5 pentads (groups of 5) of power, with a total of 24 helpers. In turn, in these doors and helpers there are 9 enneads (groups of 9) in each one, 10 dékadas (groups of 10), and 12 dodékads and 5 pentads, all of them under an Overseer, or Overseer. With him are Geradama (the great ancestor), Aphredon with his 12 benefits, Adam with his 36 aeons, and the Perfect Mind.

The great Overseer has 3 aspects: ungenerated, true and ineffable. He has his gaze in, out and up the doors. His inner gaze is towards the 'Holy of Holies', which is the Infinite head of the Sanctuary. Its outward-facing aspect observes the outer eons. Another aspect looks outward, and another looks at the Setheus, within. The Holy of Holies has 2 aspects: one opens towards the Depths, and the other opens towards the place of the Overseer, called 'Child'. And there is a Fourth Place, after this, of the Second Demiurge, which is called 'Deep'. The Deep One has 3 paternities: 1st is the hidden Father, the Hidden God; 2nd is the Father who stands on 5 trees with a table between them, and who has the 12 aspects of the Mind of the All; 3rd is the Father in silence, the source that is Love, the Mind of Everything and its 5 seals. Behind these paternities is the Mother, who is manifest in the following ennead: Protia, Pandia, Pangenia, Doxophania, Doxogenia, Doxokratia, Arsenogenia, Loia and Iouel. This Father and Mother are 'The First Unknowable' (Akatagnóstos), which completes a decade (the ennead of the Mother and the filiation-uniqueness of the 3 paternities) completing the monad of the Unknowable One (Agnóstos).

Then there is a Fifth Place, where there is hidden a great wealth that supplies the All. It is the Immeasurable Depth. It is a table that gathers 3 greatness: A Still One, An Unknowable One and An Infinite One. In these 3 there is a filiation called 'Christ Verified', and it has 12 aspects, where there is an Invisible Aspect, an Incomprehensible Aspect, an Ineffable Aspect, a Simple Aspect, an Imperishable Aspect, an Impassible Aspect, an Unbegotten Aspect and a Pure Aspect. This place has 12 sources called 'Rationals', which are full of eternal life, also abysses and 12 spaces that encompass everything. Then there is a Sixth Place, which is the Setheus depth, which is within everyone. Setheus is surrounded by 12 paternities, each with 3 aspects, making a total of 36, and apart from her, 12 surround her head and she has a diadem that radiates all the worlds with rays. The 12 fatherhoods and their aspects are the following: the First Father is indivisible, with an infinite, invisible and ineffable aspect; the Second Father has an incomprehensible, impossible and flawless appearance; the Third has an unknowable aspect, of incorruption and of Aphredon; the room has silence, source and impregnable; the fifth without aspect, with an almighty and unbegotten aspect; the sixth has everything-father, self-father and progenitor; the seventh of total mystery, total wise and of all sources; the eighth light, of rest and resurrection; the ninth covered, of a first visible aspect and of self-begotten; the tenth of three times masculine, of Adamas and pure; the eleventh triple power, perfect and sphinter (spark of light, spark aspect); and the twelfth of truth, foresight and thought.

It is funny that to relate all this, only one word from a book dating back 3,500 years had been told to us: Barashit. No? That was "the beginning", "the beginning", or in Persian, "heaven". You can complement this thesis by reviewing Sakla I, because shortly we will return to the story that they told us, not issuing judgments, but understanding that these things could not be understood until Yeshua was manifested as Master on Earth, and because, in addition to all this, only he knew them up to that moment. He remembered all his previous lives, and not only those lived in the Corpor worlds, but those experienced in the Atmos (Psyki), Es and Ethe spheres. By the way, I will call reality 'Es' as it is done in Hebrew: Ruaj (this is so as not to confuse us with the suffix 'es', in the sense of patronymic or locative nouns to form demonyms such as origin or provenance).

So, we now find ourselves with a precursor state of Genesis, the possible motor that would explain a creation as a result, not as a root (as my father, Rabbi Félix G. Van Katz, theorized in his theses for more than twenty years). An important component emerges from certain sources, such as 'The Gospel of Truth', which expresses that the 'Father', or sphere of the 10, wanted to "know Himself", and that is why the Will existed, since otherwise no one in his uniqueness would have had Freedom. The separation came from this Collective Mind that was in the Ethe of the lower realities. Since some things are replicas of others on a larger scale, everything was replicated on the scale below it. In this way the 'Aham' (Everything, in Sanskrit language) was the origin of Adam who is called Elohim, or god, which also means Man or son of Man, that is, son of God.

It is, then, that from the eons of the Invisible One came the Primordial Father, who is also called 'Father of the Universe'. He who is the Master of the Universe is whose race is said to be the kingdomless generation. Yes, the multitude that comes from Him are called 'Sons of the Uncreated Father', who have no kingdom over them. It is also known as 'God', 'Savior' or 'Son of God', since everything is a replica of one another in each aeon, ogdoad, universe and world. For this reason, Adama and his kingdom are in our likeness, and we are in theirs, as Solomon expressed in the so-called "Star of David." There is the Geradama, who is the ancestor (although Geradama in modern Greek means "old age"). They produce by their will, and those who follow do so by their thought, those who follow by their word and those who follow by their power, until they reach creation by conception and by works of hands. As the text says, "heaven is the invisible word of the father", according to Yeshua in the 1st book of Ieou. The superior glories projected the Setite races into that universe of light. Mirotoe (or 'Meirothea'), who is the mother-womb, or womb, the manifestation of Barbelo as his generative-glory, the living power and great light from whose cloud Adamas was begotten, descended there. It was Mirothea (Meirothea, Mirotoe) who produced that Adam-Man-God of the imperishable kingdom of the Christ-Son.

The Kingdom of Uios

Just as the great luminary Plesitea is the mother of angels, mother of lights, they were the image matrices of Barbelo. She produced the Setite race "bringing the fruit from Amrah, where Sdom is, which is the fruit of the fountain of Amrah" (1st Treatise of Set, NH). The Hebrew letter Samech, from the word Sdom, seems to have a meaning of "dual need", in the sense of the ego experience that a soul must live in order to then correct itself. Plesithea produced the Setite race through the mediation of Set, who took this "fruit" and placed it in the kingdom of Daveithai (or 'Daveite'). When the generative spirit in matter, and of matter, which were Sakla and Nebro, began their creative action, the angel Hormos was sent to prepare the virgins of the "corrupted generation" of the aeon emerging from chaos a holy generation of the seed of Set. He, Seth, brought his seed, then, and it was sown in the aeons that had been produced. All this occurred through the Logos, and its mediation was through the Edocla light-consciousness. This is the race that would resist because of their knowledge of its emanation, since the Setites were divided into a conscious portion of the truth, and an unconscious portion - temporarily - of the truth.

The Father wanted to reveal his name through his Son. That's why he calls it 'Uio', which is an anagram of 'Iou', cognate of 'Ieou'. This is the secret name of the Father: 'Ieu', also pronounced 'Ieo', 'Iou', Ieou', and in several manuscripts it appears as the 5 vowels, 'Iaoeu', in several variants, where usually the 'Iota' is the first of the letters of the name. The Father established in him the production of emanations and the thousands into existence. Thus, Iou is the true god, the first god. He produced 12 emanations, which are "his twelve heads in each emanation" (1st Book of Jeu, ch. 8). His name is 'Dodeka' ('Twelve'). Somewhat more precise descriptions of these partitions of the first emanations are found in the nameless text of the Bruce Codex, if you wish to dig deeper. It is not my intention to go into great detail about the higher realms, considering that you yourself can read the manuscripts, study them, make your own diagrams and draw your own conclusions. The key here is to understand why this universe was created, what it was created for and what its destiny is. For this reason the name Jehovah, or Yavé, is the deformation of the Greek 'Ieu' or 'Iao', and of the Hebrew 'IHVH', which would be 'Iaheveh', but in essence it is a game of the first person of the staff in the past tense, present and future, as a concept of "being" what actually is, since the rest is not real.

When the Christ, who is the Self-Generated god, was honored and glorified, and received the 4 kingdoms and the 12 kingdoms, he received Youel and Esefec as powers from the superior light. The four luminaries appeared together with Set, all 5 of them together. This pentad of the Son then began to receive the virtues from above, from the Ethe of the superior Ogdodada, from the One Father, towards these realms of the second reality within the Pleroma. Then came the virtues: Grace is in Harmozel, sensitivity in Oriel, intelligence in Daveithai, and prudence in Elelet. With Set this is the first ogdoad of the self-begotten divine, Christ. Later assistants came: Gamaliel for Harmozel, Gabriel with Oroiel, Samio with Daveithai, and Abrasax with Elelet. These 4 are the ministers of the 4 luminaries. Each of these four is accompanied by power, he being one power, and being with two more powers, and each of them existing for an aeon produced for the ogdoate of the Son. Thus the first three of the first aeon are Armedon, Nousanios, and Harmozel; of the second aeon are Phaionios, Ainios and Oroiael; of the third Mellephaneus, Loiios and Daveithai; of the fourth aeon Mousanios, Amethes and Elelet. This is the great aeon of the Son, and thus the invisible 24 is completed: the 3 aeons (One Perfect Father, Mother Barbelo and Son Christ), which are the great ogdoates (8 sub-eons) of the Pleroma. This is also called 'The 24 Mystery', or 'Last Mystery'.

Elelet is the star of wisdom and intelligence, or simply Elelet is wisdom and intelligence itself. It is one of the 4 luminaries that are upright in front of the Great Invisible Spirit. However, he is Sofia, and his Sofia called Pistis (Faith) wanted to produce a work by herself. He is the king, and exercises dominion over this aeon, whose lowest part of matter/earth and chaos is the demon Yaldabaot (Ialdabaot, Samael, Sakla). Sofia is the paredra of the Father-Adamas, and therefore, of the Christ, all creating together. Sofia is the syzygy that came out of the self-generated, the Immortal Man Adama. This Adam is the so-called 'God of gods' and 'King of kings', the Primordial Genitor or Man, intelligence fully understood by itself, and likewise the Protophanes, or Invisible Mind. This Adam also created an aeon called ogdoado, as an image of the superior in the sphere of Silence. Ergo, after this the Genitor and Sofia (Wisdom) meditated together and declared their son androgynous. In this way the Christ-Son of this totality has a masculine name and a feminine name. The masculine is 'First Genitor Son of God', which in Greek is 'Protogenos Uios Theos' and in Hebrew 'Itzor Rishon Ben Elohim', also defined in the Kabalah as the Adam Kadmon. Her feminine name is 'First Generating Wisdom, Mother of the Universe', which in Greek is Sofias Protogenos Mitera Símpantos (some might read 'Kosmon' (world) instead of 'Sympanthos'). In Hebrew it would be 'Jokma Itzar Rishona Am-Olam'. She is also called 'Love'.

This androgyne is the firstborn called Christ. His kingdom is the kingdom of Adam, or 'Son of Man', which is the same as 'Son of God', which in Hebrew means 'Elohim'. According to Sophias, this kingdom made appear a great androgynous light, whose masculine name is 'Savior', Genitor of All. This in Hebrew is 'Mashiach, Itzer HaKol', and in Greek it is 'Sotiras, Pantogenetor'. Her female name is 'Wisdom', Generator of All, which in Greek is 'Sofias Pantokratora', and in Hebrew is 'Jokma Itzeret HaKol'. She is also known as 'Pistis', which in Spanish translates as 'Faith', and in Hebrew becomes 'Emunat'. Two aeons have been produced from the 'Savior': that of the son of man and that of the Man Adam. Around these two aeons there is a great aeon that has no kingdom over it. This is the assembly of the 3 aeons whose totality is called the 'Assembly of the Totality', and it is also androgynous, having a masculine and a feminine name. His masculine aspect is called 'Assembly', which in Greek is 'Synarmologisi' or 'Synagoge' (synagogue), and in Hebrew is 'Keilah', his feminine aspect is called 'Life', which in Greek is 'Zoi', and in Hebrew it is 'Jai' or 'Jiah', since it is the First Eve (Eva in Greek is Zoe, and in Hebrew it is Javah, from the voice Jivah).

THE VEILS

This Jehovah produced the gods, from her they were revealed and so called. Therefore, this Assembly Elohim revealed by his Wisdom gods who revealed gods, and those gods revealed lords who, by their thought revealed lords, and those lords, by their power, revealed archangels, who, by their word, revealed angels, and Through them, similarities were revealed, with their structures and shapes. It must be understood that when one speaks of "gods" one speaks of the Hebrew definition 'Elohim', and when one speaks of "lords" one speaks of the Hebrew definition 'Adonim', from the singular 'Adonai'. Yes, many Fathers, many Adams, many Wisdoms, many Elohim, many Christs, many sons, many Seths, many Adonai, and besides all this, all of them are ONE. These gods, lords, archangels, angels and similarities that come from the source receive their authority from the Immortal Adam and his Sophia called Silence (Siopi). Sofia, the Mother of the Universe and the wall, wanted to bring the chosen ones (the Setites) by herself - without her counterpart - so the Father of the Universe created a cloak or veil called 'Spirit', that is to say 'Ruaj' in the language Hebrew, or 'Pneuma' in Greek, which also mean 'wind' (a conscious but invisible movement), separating the higher realities from the lower ones. So that there was a veil of the Ethe realities, a veil of the spiritual realities and a veil of the psychic realities, and under all of this matter.

If you have ever seen the distribution of the temple of Ierushalim (Jerusalem), you will understand what I am talking about. There you can see how there are various partitions, such as the exterior of exteriors, which is called 'Jetzer ha.Goim', or court of the Gentiles. What does this mean? The 'Goi' is the world-cosmos, and the 'Jetzer' is the expanse in which it is found. The courtyard is the same as the atrium, whether enclosed or open, representing space, and the Goi is matter, the composition of the physical worlds. The 'Jetzer ha.Pnimit', or Inner Court, was the part from which the original structure began. What does this mean? In the original plan there was no such 'Jetzer ha.Goim', as this is a reflection that in the "original plan" there was no world outside of unity, whose symbol is the Temple. Some interpreted that the Temple symbolized the body as the temple of God, but in reality the symbolism went much further, since it was not intended to legitimize an illusion, but to teach the truth about the Monad and the Pleroma.

In this way the Jetzer ha.Pnimit are the outer realities of infinite space, where there was only consciousness, not matter. Then there was the Temple, which had a first veil towards "the external realities", which are the veil towards the Ethe. As we see in the number of objects in the temple and their distribution, this is how the exterior of the interior realities of the ONE is expressed, which is covered by another interior veil. For this reason it is called 'Kodesh' (Holy) and inside 'Kodesh ha.Kodashim' (Holy of Holies), which in Latin was called Sancta Sanctorum. That is the kingdom of the One, its monad, where, moreover, is the 'Aron'. Why was Moses' brother named 'Aaron'? You see. Aron is 261, like 'Ha.Dbarim' (The Words), symbolizing the source of all things. Inside are the 3 symbols: of the Torah (instruction), Power and Knowledge. That is the true Mikdash (temple), which means 'my kodesh' (of the holy, where the sacred resides), whose number is 57, like 'Shamaim' (heavens). Now let us leave the Holy of Holies and remember what is outside the inner veil: shew tables with loaves and candlesticks. What do they symbolize? The kingdoms of the son. The loaves symbolize the Son of Man, Adam; the tables, the supports of the kingdoms; the candlesticks are the imperishable kingdoms, and the lights are their stars-angels-princes. From them we leave by crossing the veil towards the homeless realities, since they are unprotected and exposed, vulnerable. For this reason it is crossed by a Marpeset (porch) of two Amud (pillar or column): duality. The Mazbaj (altar) is lined up there, a few meters in front of the door.

Let us note that the word Amud is numerically 39, like 'Ben ha.Adam' (the son of man), Ha.Shem (the Name), Kerub (cherub), Rafah (healing), Zait (olive tree), Midbar (desert), Laila (night) and Aretz (earth). And Amu d is likewise 120, like the phrase from Genesis, which says, " Zajar ve.Nekebah Baram," which translated is, "male and female were created." That is, because each column symbolizes an aspect of duality: masculine and feminine. The separation of the mind, of the oneness, into two different parts. That is what the concept of husband and wife means, where one is Christ and the other the New Jerusalem, that is, the elect, which is the same as the called, or the virgin. What should I talk about? Of the return to oneness, whose process is the entrance to the bridal chamber with the husband, that is, the spirit, since only "in the secret place", inside, is there the truth. For this reason, in Hebrew mysticism, 120 represents the end of the flesh. Furthermore, the Amudim are in the Marpeset (porch or porch), which numerically is 87, as 'Melki-Tzedek', who is the companion of Iao as guardians of the portal to the Kingdom of Light and Justice (Melki-Tzedek means ' king of Justice'). 87 is also the number of Abodah (ministry, work) and of the phase 'Ani IHVH' (I am Yaheveh), as of the Greek

'Parthenos' (virgin). By Marpeset it is also 580, as 'Atik' (ancient), in reference to the Ancient of Days.

For its part, the altar identifies Pistis, or Love, as we see in the numbering of Mazbaj (30), which is the same as the Greek 'Agápi' (love). That is the altar (Mzbaj), the area of "sacrifices" (Zbaj), which bears the 'M' of Matter, of the place. The Mikdash (Temple) is also representative of the Mind: unconscious (Kodesh ha.Kodashim), subconscious (Kodesh) and conscious (Jetzer). Consequently, at whatever level we wish to see it, this is the higher realms of those produced by Sophia the Little, who is also called Pistis. When she thought of doing this, a light was revealed, like a drop of light that came towards the lower regions of the omnipotence of chaos, and through the spirit she revealed the modeled Forms, giving them with her breath (Nishmat) a living soul (Nefesh). Jai). I will explain this last section in more detail in the chapter on Day Eight.

The Projection

Space already existed before the Big Bang, and would be the lower part of the veil that separates higher realities from lower ones. The Big Bang would only be the projection of the dream into space, or the mind. But the Big Bang is not the veil itself. The Ruach-Spirit is the veil. The upper part is the pure Ethe reality, and the lower part is the structure of sub-Ethe projections derived by the Logos in the states Atmos-Psyki and Corpor (Material), whose generic concept is called in the East 'Maia' and 'Anicca'. '. There was, again, a space, and in it an idea of chaos. This is because first there was an idea of individuality in the Elohim Collective Mind, that idea of individuality created an idea of separation, and that idea of separation produced an 'I', or 'Ich' (in the German language, but pronounced 'Ij').), which is Self as an individual and separate identity. That 'I' - in Greek 'Egó' - then a personification that could not be lacking in power, given its potential nature as Elohim. How could you illustrate this? One way would be to imagine Infinity as the unlimited expansion of space, and the Big Bang as an explosion within Infinity. That explosion occurs in a parallel space of infinity.

That explosion projected duality, and that explosion came from the pristine source. This is what Vedanta refers to as Brahma created all things, but from that source came the gods Vishnu (good) and Shiva (evil), the principles of the polarities of the dual universe. They are not the source. The source is Brahma, but Maia's dream and all its dimensions and planes depend on their projection. String Theory tells us that dimensions are intertwined, with overlapping realities, parallel universes and intersecting planes. In this way, in what might seem to be the same location, there are multiple intertwined realities. Therefore, "the Kingdom of the Father is spread over the earth and people do not see it." (Ev. Thomas 113/112)

What apparently is nothing, is, in reality, a cosmos of physical matter condensed into fields, crossed with multiple states. There are linked and traversed universes, planes, dimensions, realities and spacio-temporal continuities. These spheres of reality would not be noticeable to a greater or lesser extent because they fluctuate in different vibratory states, foreign to the perception of the consciousnesses that are experiencing at a frequency of lower wave oscillation. That means that the waves do not collapse between them but rather cross and overlap. In this sense, states of randomness and atomic and molecular organization would have been linked. Random states, or Chaos, would not conflict with Ethe states, since Ethe sustains all realities, whether spiritual, psychic or physical. The Ethe is the substance of the higher realms. These kingdoms are divided by a veil, and under the veil there is a tiered system of levels of the aeon produced by Sophia the little one, that is, by Ahab (Love), whom in Greek they call Agapi. Veil in Hebrew says 'Paroket' whose numbering is 70, the same as 'Sophia' in Greek, and wine, both in Greek (Oinon) and Hebrew (Iain).

That veil is traversed in a state of vibration of consciousness that is already ONE with the Infinite. That veil is crossed when overcoming the eighth of this aeon, where the satellites, or Probolos, of light are, which is the kingdom of Justice. It is a state that transcends the 12 sub-aeons of this material eon, beyond Chaos, and beyond the 13th sub-aeon, where the 12 invisible ones are, the probolos. Those are the places of the Mysteries. There is therefore, between the Region of Justice and the lower sub-aeons, a region called the Middle, or the Center, since it has Justice on the Right, and injustice on the Left. Those are the upper regions under the veil, which are the bridge, or Bifröst, to the 'Ein Sof Aor'. But getting to that requires the path through a scale of 7 states of vibrations or densities of consciousness.

Day Zero

As the Gospel of Truth affirms, "the deficiency occurred because the Father was ignored" (vers. 24). The First Treaty of Set says that 5,000 years after the creation of the Setite race, someone was sent to govern Chaos and the Invisible (bottomless, which in Greek is called 'Adis' (Hades)), and there the angel appeared Sakla with her demon Nebro, to become both a single "generating spirit of matter". When the Collective Mind of Elohim thought of the separation, a parallel reality was produced, projected as a White Hole. As the Mind of Elohim produced the idea of separation, the processes that configured this idea condensed into focalizations of consciousness. However, while the focalization was taking place, the random states were chaotic, and the matter that was then consolidating, but not yet forming a functional structure, was distributed as an abyss of gaseous matter. All this was moving through the existing Ethe worlds, creating an expansion, which science considers that a posteriori would be collected to return everything to the starting point.

Now, the potential of the idea of separation was created as a dual power, embodying the non-luminous polarity of the Yin Yang conception. This ego of the Collective Mind would take form as a matter that would create a sort of multi-level realities, from divine duals to matter. This dual ego principle that took form became self-aware, and saw itself as a high-level power of density-dimensionality, comprehending part of a reality: it was a god. To the imperishable immortals, he was but an angel, but to him and his reality, he was a god, and not just any god, but a jealous god. This being arose from the potential of material creation as a lion of fire, which is why it was called Ariel (which in the Hebrew language means 'Lion Deity'). Duality was evident in him as two aspects: Sakla and Nebro. Sakla is the demiurge of matter, and has 3 names: Sakla is his foolish aspect (from the Hebrew voice Kesil, which in Aramaic is Sakal), Samael, which means 'Blind God', and Ialdabaot, which means 'Born of the Abysses'. The darkest aspect of these two beings is Nebro, or Nebruel, which is their stupidity, the demon with whom they both created a spirit that generates life and matter in this cosmos. These created by replica of the things above a whole pantheon of authorities-principalities, powers-powers, powers, kosmokrators, kings-gods, archangels, demons, angels and virtues, and these produced bodies in the worlds they ruled.

They prepared their cohort to prevent the Setite race from coming to take power from them. That side of dark polarity is known as 'the Left', in contrast to the powers of light - called 'the Right' -, which is the existence that belongs to the Kalyptos, that is, the true Divinity. In Jewish tradition it is said that Adam had a wife before Jevah, and they call her 'Lilit' after the phoneme Lila, from which the sound Lailah (night) comes. Although I also explained this previously in 'RS2', I did not address its metaphorical meaning. Lilit is the idea that attempts to allege that the mind conceived an ego before the separation was projected, that is, before it materialized. This symbolizes the kingdom of Elelet, where the possibility was conceived, or put more concisely, the state and projection where those who at first would not understand the Fullness, but would ultimately regret it, would manifest. In Jewish tradition Lilit becomes Samael's consort when she is rejected by Adam. This is because not all Mind conceived duality, not all believed in the ego, that is, in separation.

Although, the Mind "thought" of the idea of separation, but it was not accepted by the entire kingdom of immortals. Therefore, it received life in a portion of the Mind, which was its unconscious aspect, and from it this eon emanated (the cosmos called 'World' in the Spanish Bible). However, since the idea was not typical of the Right Mind, it only had a place with the personification of the ego, which was the emanation of dualism, whom Samael - or 'Ialdabaot' in the texts of Gnosticism - identifies. It therefore makes sense that Jewish tradition affirms that Samael and Lilit conceived the lilim, shedim and ruchot, as well as Ashmadaeva (or 'Asmodeus'), who is the demon-angel of wrath. It is also logical that in that same tradition the creation of Samael-Lilit is compared with that of Adam-Jevah, arguing that the former were the masculine and feminine of the same reality produced simultaneously. However, having projected all that multitude of ethereal lights into the dream, they configured through consciousness a spiritual body for each apparent individualization, called in Hebrew 'Nefesh', in Greek 'Psiji', in demotic 'Ba', in Sanskrit 'Atman'., in Latin 'Anima', in English 'Soul', and in Spanish 'alma'.

Ergo, through many emissaries the protection and guidance of the Setites was prepared. Through the great Set came: the great auxiliary Yeseo Mazareo Yesedeceo, the living water (water of life); the great guides Santiago the great and Teopempto and Isavel; those who preside over the fountain of truth, Micaeus, Mikar and Mnesinus; the one who presides over the baptism of the living and the purifiers, Sosengenfaranges; those who preside over the water gates, Miqueo and Mikar; those who preside over the mountain, Seldao and Eleno; the greeters of the race of the great Set; the 4 ministers: Gamaliel, Gabriel, Samio and Abrasax; those who preside over the Sun, its birth, Olses and Hypneo; Eumario and those who preside over the rest of eternal life; Mixander and Micanor; those who keep the souls of the chosen one, Acramas and Stremsujo; the great power Heli Heli Majar Majar Seth; the great virginal spirit; the great luminary Harmozel, and Adamas, who is with him, who is the first place; Oroiel, which is second place; Yeshua (Jesus) of the third place, which is with Daveithai and the race of Seth; the place where the children rest, which is Eleleth, as a room; from fifth Youel (Ioel), who presides over the name – or secret identity/purpose – of the one who would be allowed to baptize in the heaven-exceeding holy baptism.

Part II

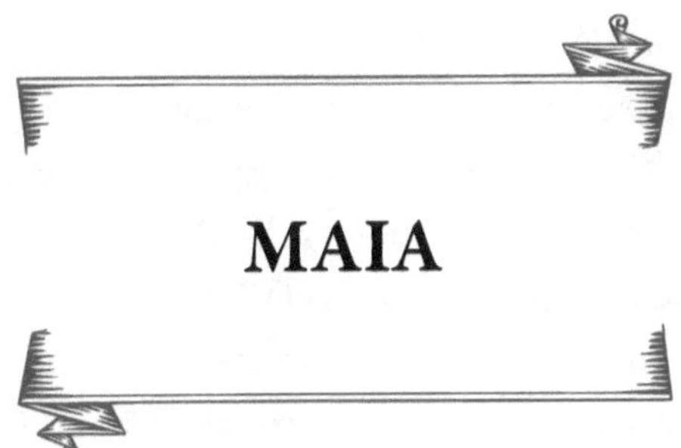

MAIA

THE ILLUSORY MATTER

The Kabalah defines 4 states of reality, similar to the 4 consistencies (Ein, Ethe, Psiki-Atmos and Corpor), which they call Atzilut, Briah, Yetzirah and Asiah. As I told you before, in Jewish mysticism the 'Ein' is "He Who Is", and that is Nothing, but in reality Everything, and from Him derives 'Ein Sof' (No End), which is the all-encompassing consciousness, and from Her the 'Ein Sof Aur', as unlimited light and conscious totality in infinite light. The conception of emanations and ideas would derive from Him, since they come from the Mind of the One, where His Thought resides. This principle, called 'Atzilut', is represented with the Hebrew letter ' **Yud** ', and with the element fire. That sphere of emanations are the pristine aeons of the Christ, and from there the creations were produced, which in Kabalah they call 'Briah', which is the level of the images, and which they identify with the letter 'H e', and **with** the water element. You will then see how the two subsequent levels fit together, which are Yetzirah, linked to the letter ' **V** av' and the

70

element of air, and Asiah, linked to the letter ' **H** e' and the element of earth. This is so because in Hasidist mysticism the concepts of the Kabbalistic Tree of Life are an exemplification of the descending levels from the source of all sources. With this representation they put together the tetragrammaton: IHVH.

" [Matter engendered] a passion devoid of similarity, since it proceeded from an act against nature. Then there is a disturbance in the whole body." (Ev. Mary Magdalene, verse 7) Body? Yes, organism, at any level. The universe is a body. Maia is said to be the name of the Illusion, which others have called Anicca. The letter 'M' is that referent on the matter, not being a coincidence that Yeshua had a biological mother named Mariam, and a wife also named Mariam, as Buddha had a mother named Maia. Precisely Maria in Russian is said 'Maia', although, although the name Mariam passed to Maria, in Hebrew it is written only with the characters Mem, Reish, Yud and Mem, that is, as 'Mrim', or 'Marim'. It should be noted the similarity of the phoneme 'Mar' with the cognate of Mariam. This is because, regardless of the fact that Mariam means 'rebellious people' in Hebrew, it has the particle 'MR' which in demotic (ancient Egyptian) means 'pyramid', and the phoneme 'IM' which means 'Sea'. The Hebrew form 'MR', can sound like 'Mir' (since this is where the optional sound structure 'Miriam' derives) or 'Mor', which means "myrrh". Precisely in Hebrew, 'MaR' means "drop of water". However, read in another way, 'Mariam' comes to mean "rebellion of the sea", since 'Marei' is rebellion, and the letter 'M' alone symbolizes water. On the other hand, the particle 'Ramah', is to be thrown down, while 'Ram' is to rise up, or "wild bull".

In short, she symbolizes what Christian theology calls "The Rebellion", but not of Lucifer, but the separation of the Son of Man from the original source. Therefore, Mariam herself represents the projection of the return path. It is curious that in our languages we use so many definitions without understanding that "words" are structures of letters, which form syllables and sets of sounds. Sounds are articulations of vibration in the air that have meanings of power, as they come from the mind that devised them before uttering them in undulations of various frequencies. All the Om, or initial sounds are actually potentials that reflect thoughts. Syntaxes, for their part, are reinforced ways of trying to transmit in structured sounds concepts that belong to a World of Ideas. This is how the apostle Philip expresses it when he writes: " The names given (to things) in the world are susceptible to great deception, since they distract attention from what is stable (and direct it) towards what is unstable. And so whoever hears (the word) "God" understands not what is stable, but what is unstable. The same thing happens with the "Father", the "Son", the "Holy Spirit", the "Life", the "Light", the "Resurrection", the "Church" and many others: the (concepts) are not understood. stable, but the unstable ones, unless the former are known (in advance). These are in the world [...]; If [they were] in the aeon, they would never be named in the world nor cast among earthly things; they have their end in the eon." (Ev. Felipe, verse 11)

So what's the point with Maria? Maria, Mariam, Miriam or simply Maia, is a sound that reflects the identity of the 'Mother'. Every archetype exists in the world, as Carl Jung expressed, on the idea of a primordial mother. There is, then, a primal mother, where she is known as the unmanifest and the manifest; but there is also a mother who is later, who is known as the one who has not been revealed and the one who has been revealed. The one not revealed was the great Sofia, but the one that was revealed was the little Sofia, also known as Zoe, Vida or Eva. Philip wrote: "Some say that Mary conceived by the power of the Holy Spirit: these are mistaken, they do not know what they are saying. When has a woman ever conceived as a woman? Mary is the virgin whom no Power has stained. She is a great anathema to the Jews, who are the apostles and the apostolics. This virgin that no Power has violated, [...while] the Powers defiled themselves." (Ev. Felipe, ver. 17)

that the names Sofia, Wisdom, Zoe, Eve, Pistis, Faith, Love, Life, **Mary, Maia, Matter, Mother, Mass, Maim** or Miriam define the same thing: a Mother. The true Mother is a virgin, as is the figure of the constellation Virgo. Consequently, what was Felipe referring to? The Holy Spirit is the Bosom of the Father, who is manifesting within the illusion of sleep (Maia-Anicca) through forgiveness. Sofia represents our Collective Mind, because it is greatly wise. It (our mind) has unconscious guilt for thinking that it has separated itself from oneness. The estrangement from filiation is not real. The dream (whose beginning science calls 'Big Bang') is unreal, and has only occurred in the Mind, that is, by an idea of Sofia the Minor. We believe that we are separate from God, but it is absurd considering that we are God, part of God, we are within God, and God is within ourselves.

We are God in the idea of separation as in the indivisible reality of waking, which is paradise. That is why the passage from the apostle John says that "God so loved the world that he sent his only begotten son." Who is the only begotten son? Unigénito in Spanish means "only generated." If it were Yeshua alone, we would not be children of God. If only "one" is the Son, however, we are part of that Son. It is because of 'Love', that is, of Pistis, that from Elohim the son of man came to the projection of the dream, that which some call World and others call Cosmos, that is, Maia-Anicca. Sophia is completely a virgin, and she is the one who conceived an only begotten Son, not the daughter of Anna and Joakim, wife of Joseph the architect of Natzareth. That is why Philip says in verse 32 of his gospel, that "Mary is, in fact, his sister, his mother and his companion." The Mother is the wall and sister of Elohim, who is also the incorruptible Father, the Adama (Man) and the Verified Christ, for all these are one and the same. Mary is his mother, because she is the womb and at the same time the midwife, and she is the companion, because with her he conceived everything.

Why wasn't Mariam - daughter of Joakim and Ana - pregnant by a man? She gave birth alone, because it represented the separation that existed at the beginning, according to the decision of Pistis, which was to listen to the devil, that is, the ego (as seen in the Greek myth of Rhea-Juno creating Hephaistos-Vulcan). This has been referred to by Moses in the mythological metaphor that says that "Eve" was tempted by a demon. But Judaism says that Eve became pregnant and gave birth to Cain. It is then that Pistis did conceive with someone, but it was not the Christ, but the ego, since the idea of separation cannot help but create the unlimited potential being. The Mind received an unlimited potential, do you remember that I told you in previous pages? Kain means "the jealous one," as it represents Ialdabaot, and Abel means "vanity," as it identifies the cosmos. But when Maia was produced the idea was replicated, since now the repetition of the fractal sequence projected Ialdabaot raping the minor Sophia, since she is also called Life, that is, Zoi, which in Hebrew is Jivah, or Eve. Thus he created his powers, which are those of the jealous god, the Kain. For this reason Pistis also provided the race of gods of justice within this aeon, who are those of the side of Iaheveh, who at first due to their innocence were prey to the madness of the archons.

Some see the Holy Spirit as a mother, as when Yeshua said, "even so did my mother, the Holy Spirit, take me by one of my hairs and lead me to the great mountain Tabor." (Ev. Hebrews, fragment of Origen, Commentary on John 2.12.87) The apostles referred to their working portion as Melach (salt), so Philip defines the Wisdom as such. As I try to make you understand, the sounds that words produce, and their names, reflect thought potentials. Melaj is salt, according to the Hebrew letters Mem, Lamed and Jet, and by phonetics and cognate we find the relationship with Melej (king) and Malaaj (angel). The difference between king (MLK) and angel (MLAK) is an Alef that symbolizes the divine. These all start with Mem, and then Lamed, as a connecting representation of the higher and the lower. As for salt, the final letter denotes life, even though salt itself is not supposed to have it. Salt is made up of sodium chloride crystals that grow in combination with oxygen. Analyze this sentence: "The apostles said to the disciples: «that all our offering seek salt for itself». They called [the Sophia] «salt», (because) without her no offering [is] acceptable. Sofia is sterile, [without] child(ren); that is why it is called [also] "salt"." (Ev. Felipe, vers. 35-36)

The Death of Matter

The phrase, "you are the salt of the world," alludes to this, salt being the life in matter through the work of Sophia the Lesser, who produced this cosmos (for World is a Latin word that in Greek is Kosmon). That is why Yeshua said about this world: "A vineyard has been planted outside the Father, but because it was not strong, it will be eradicated and will wither." (Ev. Thomas, saying 40) Thanks to these great truths we can understand Philip when he wrote: "While Eve was [within Adam] there was no death, but when she was separated [from him] death came. When it returns and he accepts it, death will cease to exist." (Ev. Felipe, verse 71) The world, as the dictionary explains, is "the set of created things," not our planet, as it is commonly interpreted; death is separation; Salt is what gives reason to existence; the "vine" is the material universe; that 'Adam' is the God, he is Elohim, he is the imperishable Christ of the ethereal mansions. Therein lies the mystery of Moses' metaphor, when he said in the myth that there was a "deep sleep" in Adama, since he fell into self-deception. Having deceived himself, he separated. A part of him created a difference in what until now was complete.

The life drawn from Adam-God-Christ is nothing other than the manifestation of separation. That is why in Hebrew death is

'Met', from which comes the sound 'Mevet', from the Canaanite 'Mot' (dead). The 'Mem' here is found next to the last letter of the alphabet - the 'Tau' - because death has an end. It is not the end, but it will be finished. What thing? The matter. It is then that Matter had a manifestation, but from the beginning it is destined to end and dissolve, since everything must return to the origin. The woman (Eve-Life) separated from the true Man as his image, but never returned to him, at least not in reality, not while they were still in paradise, that is, in the Ethereal Abodes. As Philip says: "If the woman had not separated from the man, she would not have died with him. Their separation became the beginning of death. That is why Christ came, to annul the separation that existed from the beginning, to unite both and to give life to those who had died in the separation and unite them again. Well, the woman is united with her husband in the bridal chamber and all those who have been united in said chamber will not be separated again. "That is why Eve separated from Adam, because she had not united with him in the bridal chamber." (Ev. Felipe, ver. 78-79)

For this reason, Yeshua explains that as long as everything was androgynous it was complete, but the separation produced duality, which is scarcity, lack, dependence, necessity, that is, incompleteness, and therefore, everything must return to the root, whose symbol is marriage, and the concept of the 'Bridal Chamber'. In the Gospel of Thomas he refers to it in the following way: "When you make the two one, and when you make the internal as the external and the external as the internal, and the higher as the lower, and when you make the man and the woman only one, so that the man is not masculine and the woman is not feminine... then you will enter the Kingdom." (Saying 22) Felipe talks a lot about the idea of the Bridal Chamber, but it is an area that I will address little, since I have already discussed this topic on other occasions - especially in my book 'Naked Sex' -. Even so, it is good that as you read these lines you meditate on the projections that manifest in the material world in polarities, and how males are dependent on females and females on males in the likeness of lack due to separation, and how They eagerly seek to unite again (even if it is for a while, knowing that no matter how much rubbing and contact occurs, they will not be one, since true union will only occur when we are in God, since it is with Him that we want to return, and the climax of the sexual act is just a glimpse of the pleasure of being in heaven).

That is the stumbling block for religious and legalists when their ego longs for a dual God, vengeful, wrathful, punishing, furious, implacable, and other characteristics completely foreign to those called by Shaulo of Tarsus, "Fruits of the Spirit." It is by studying the Mind that the Holy Scriptures are understood, since the Bible does not intend to promote dualism and the projection of illusion, but rather to explain the symbols of the ego and the Right Mind, one in front of the other. Therefore, the passages about the future "destruction", "death" and "end" of the polarities that we define as dark or negative are rooted in ego projections, not in people. It is very simple, if a being could destroy an entire generation of criminals, criminality would not disappear, since it is the Mind that is "criminalized." Soon criminals would appear again. Reasoning is understood at the level of the ego as its desire for self-destruction for the being, since it makes us think that we are unworthy, and therefore, liable to be destroyed. The Bible says that "the last enemy to be destroyed will be death," but how is death killed? Isn't that stupid? When I was young I hadn't even questioned it. The death that is eradicated is separation, for we will return to God.

Monstrous Archetypes

In the religious and superstitious world, evil entities and reptilian and amphibian creative beings have been believed since very remote eras. These archetypes are a projection of the Collective Mind as personifications of its ego on multiple levels. As I explain in

RS2, various ideas of demons and devils, as well as primal gods, are an important part of the worldview of the people of Earth. In Christian theology basically all this guilt that is sought outside is projected onto Lucifer, who they say was a supposed cherub who rebelled, and God allows him to tempt humans until one day he will be punished. In addition to being crazy, this idea of believing that God will let some being tempt us transcends very simple aspects, such as the fact that God's angels do not reveal themselves, since the ideas of duality that emerged within this eon were dissolved almost as quickly as they were produced. (the only ones that remain are those that are part of a hierarchy of darkness typical of the aeon, but not of God). Thus, there have been multiple egoic projections, or SAS (Service To Self) consciousnesses, rather than SAO (Service To Others), before Lucifer, perhaps even before Ialdabaot and his minions.

These beings rebelled, letting it be understood that the realities manifest in the lower kingdoms, and that they are from this eon, are not the first, but that there were powers and kingdoms that existed before - or parallel - to Ialdabaot. The realities and consciousnesses of other higher dimensions are more than what is perceived. Humans in the 21st century measure that perception by energy and waves. Science says that the universe is between 70% and 73% Dark Energy. The other small portion is Dark Matter, which would be 25%. Thus, matter, as such - or the "physical" (baryonic matter) - only corresponds to a tiny proportion of less than 5%, and another even smaller portion would correspond to the mass of neutrinos (uncharged subatomic particles)..

Dark Matter does not emit any type of electromagnetic radiation. Essentially without light - as we'll talk about later - perception, reality or experience in this cosmos would be meaningless to us as we understand it right now. Using a comparison with the Kabalah, this expression of the emanations of Atzilut comes to be the Briah, but it is developed by the consciousness of the

Collective Mind of Adama-God, that is, of the Elohim called Adam Kadmon, who becomes conscious in a way random in space. Consciousness would be like the perception of Dark Energy, while non-consciousness is the state of Dark Matter, and the focalization to produce Briah is the toroidal power that configures baryonic matter, or common matter. This state that the Kabalah defines as "thoughts" - which is the "Spiritual" - began with various models of dualism and its characteristics, in a series of ideas or archetypal concepts that I am going to introduce before continuing with the genesis story.

The Levels of Mind

Jewish mysticism considers the projection as a tree, which it calls the 'Tree of Life'. This is crucial, because the Mind is understood as a tree (beware! I always refer to the Mind, not the brain). The Kabalah considers that the roots of said tree are Atzilut (state of emanations), where the Elohim is projected towards the first idea as consciousness in the process of self-consciousness - which is the Crown (Keter) - the potential energy that explodes, and that we could compare in a worldview with the Big Bang. This idea in the Mind is the propagation of the ideas of the dual conception. The next stage is Chochmah, or Wisdom - not necessarily associated with the personification we call Sophia, or with Pistis -, followed by Binah (Understanding). This would be the first group at the top, since the Kabalah Tree is divided into 3 sections. This is the primal potential of energy focusing.

The process of creation - or 'Briah' - is then developed, going through Daat (Knowledge) quasi imperceptible, intermittent, sometimes being and other times not being, to produce the idea of Mercy (Hesed) - or also defined as Greatness -, which would be the idea of the desire to share unconditionally. Successively, the following sefirot arrive, accompanying this Geburah (Strength), and descending to produce the central one, called Tiferet (Beauty). There is a lot to say about the sefirot, but it is not the dynamic of this book, since my intention is to give an idea of the parallels and levels of this information, how they fit together, what they are, and in what order and form they are conceived. It is about understanding the Mind – called in Greek 'Nous' -, but not from the phenomenal state, but from the comprehension above. The lower level of this archetypal tree is that of the formation - or Yetzira -, which has the spheres of Netzej (Victory), Jod (Glory) and Ysod (Foundation).

Now, the point that I really want to make is that despite the idea of three levels, there is the characteristic of an additional, 4th level. The number 9 is already infinite, but an aspect is added, which is that of Malkut - the kingdom -, the completely material world. This one has a dark perception called Qliphot, with its own 10 negative aspects. With the example of the tree, Atzilut contains the roots, Briah the trunk, Yetzirah the branches, but Asiah has the flower and the fruit. The importance of this scenography of varieties of names, colors, concepts, levels, etc., is that they are a perfect iconography of the unconscious, subconscious and conscious aspects of the Mind, and therefore, of the processes in which the Maia's dream - and, by the same routes, one returns to the source -.

Therefore, the tree develops in the way previously explained, but the path back to the source does not start from above, but from below, starting from Malchut. This is taught in Vedanta and Yogi mysticism as a snake that climbs up the spine through 7 energy sources, called Kundalini. In my work 'Naked Sex', I deal with the issues of the first energy source - or the red potential -, but here I will deal with the 7 energy sources, or better known in Sanskrit as 'Chakras'. According to Rabbi Isaac ben Jacob Ha.Cohen's treatise ('The Emanation of the Left'), Keter had 10 emanations of ancient powerful, virtuous spirits of this primordial principle. Their names, according to their order, would be: 1º the prince of the exalted heights, Sabi'El; 2nd the wonder prince of wisdom, Peli'I'El, or

Sagsagel; 3° Yerui'El, who is the prince of those who keep understanding.

These three are in the roots of the archetypal tree, and come under them, and after them, 7 emanations or degrees: 1° Memeriron, prince of Benevolence, associated with water; 2nd Geviriron, the prince of the invincible Force; 3rd Yedideron, Prince of Mercy; 4th Satriron, prince of the foundation of the world; 5th Nishiriron, the prince of the victory and triumph of Israel (the Sethite people, or church); 6th Hodiriron, the radiant prince of majesty; 7th Seforiron, the radiant prince of the last emanation of all the degrees. This pattern of 7 follows the dynamics of the 7 lights, or 7 chakras, which are the 7 levels of density, that is, the 7 dimensions of the octave. The path of the Kundalini - or ascent of the sefirotic tree back from the fruit to the root - is nothing other than the return through consciousness towards the source of the Unconscious Mind in order to be aware of it as a whole. In other words, it is the awakening of the conscience, of the Christic mind, of the conscious participation in filiation, the full union with the Holy Spirit, the dissolution of the ego. This ends after 7 dimensional levels of experiences lasting several cycles of this aeon.

When these emanations that I have just mentioned gave virtue to the kingdoms of this eon at their highest levels, when the first angelic beings were going to be produced - says rabbinic tradition - they were in a latent state, from which they would then manifest, and in In these states there were dual, discordant ideas. Of course, the Collective Mind had already produced the idea of separation, and consequently there would be various singularities of the Separation type, projections of an "egoized" mind. These emanations of angels for the aeon of Sophia the little one could be conjectured to have even arrived after Ialdabaot, yes, when Tzabaot rebelled against him and joined the light. Thus, the first ideas of angels could have been impregnated with duality and polarity, and would explain the origin of Qamti'El, who was not later manifested, but with his subordinate angels would have returned to the basis of the creative thought of Tzabaot and Pistis - who helped him in the creation of his kingdoms in his heavens -. After Qamti'El there was another deficiency: Beli'El, and the procedure was repeated. But once again, before producing pure angels, Iti'El came, and then he was dissolved in his root, in the source from which he was projected, but before being manifested. That is why the Bible says that "he saw deficiency in his angels."

The Spirit Moved

I suppose you are beginning to understand why in this work - although I should have started telling you about a biblical Genesis on the 6th day of which we talked about man - we are, already before that first day, talking about man. I will go ahead and return to this point to the extent appropriate to the story. It is key to consider that this Genesis story - and any aspect of the Bible - does not make sense in the light of the truth if that truth is not previously known and if the Mind is not understood. Don't you think so? Now, Genesis 1:2 says "go to Aretz" (and the Earth) "aitá" (was, was) "tohu va-Bohu" (abysmal and chaotic) "go to choshech" (and darkness) "to the pnei" (on face) "tehom" (abyss) "ve ruach" (and spirit, wind) "Elohim" (gods, god, strong ones) "merajefet" (fluttered) "al pne" (on face) "ha maim" (the waters). Fabulous, it tells us the state of the Aretz, but it says nothing about the state of the Shamaim. Thank goodness I took twenty pages to develop it. No? Those first words of Genesis evoke the primordial Heaven, the 'Shamai ha.Shamaim' (Heaven of the heavens), the Ethe universe that sustains the other universes derived from the Primordial Father.

The issue now is that it would lack coherence to maintain that a perfect god created something from nothing and that something was chaos. The only explanation is that this god had some type of deficiency, lack, weakness or limitation. Any analysis must be carried out always keeping in mind that both the metaphysical and physical worlds are the same; The sublunar is the image of the supralunar, and vice versa. The symbols of the spiritual and the material are not separated. Within the dream they are projections relative to each level, since the Mind creates simultaneously - despite the independent perceptions within time, or "times" -, so that each illusion or concept takes shape according to the level it is in. Therefore, what would be darkness as a phenomenon is also darkness in a metaphorical and representative sense. The idea of "darkness" as 'ignorance' is applicable in the metaphysical sense, but it does not mean that darkness did not exist in space, much less on the original liquid mass that constituted the mass from which the solar system was formed., or even more broadly in a sense of the matter itself.

An example of the understanding of symbols would be taking from the Kabalah methods of analysis of words, such as 'Joshech' (darkness), whose letters are Jet, Shin and Kaf (J-SH-K) and looking at others by phoneme, permutation or anagram. Thus we see similarity in two of the three letters with 'Najash' (copper snake): NJ-SH. If the Choshech was on the water, let's see the number of Choshech: 40. Precisely in the Hebrew language the letter of the water is Mem, whose number is 40. Therefore, it is interpreted that the darkness over the abyss was, in effect, the depth of the mass of water. Next we have the word 'Tehom' (abyss), which numerically is 46, the same as Yamim (seas). If we look at the shape of the letters of Yamim, we have: IMIM. The first Mem is open, and the final one is closed, since one symbolizes the upper waters and the other the lower, abysmal or deep waters. Then - in the same story of genesis - a curious sentence appears that maintains that "the wind of Elohim fluttered to the face of the waters".

Yeshua relates: " Then the Mother began to move from one side to the other. He realized he was missing something when the brightness of the light dimmed. She became dark because her lover had not collaborated with her. I said, 'Sir, what does it mean that she moved from one side to the other?' The Lord laughed and said, 'Do not suppose that it happened just as Moses said, "over the water." No, when she recognized the evil that had taken place and the theft that her son had committed she repented. Although in the darkness he had forgotten his ignorance, he began to feel ashamed and agitated. This agitation is the movement from one side to the other.'" (Secret Book of John 8:1-4) Why does Yeshua contradict the written version? Because as usual the shapes were visualized but not the symbols. Therefore, the same parameter of darkness and the abyss, that is, deep water, identifies matter, although not simply at the level of form and elementary atoms, but in other dimensions. In other words, structures were not only being configured on the material level, but also on the atmosphere. And what was formed in that atmosphere?

The Master explains it with this parable: "Yeshua said: 'the Divine Law of God is like a woman carrying a container full of flour. While walking along a distant road, the handle of the container broke, and flour spilled behind it along the road. She didn't know it; He didn't realize the problem. When he got home, he put the container on the floor and discovered that it was empty.'" (Ev. Tomás, saying 97) The woman is Sofia the little one, or 'Faithful Wisdom', who realized what had happened when a great power of her potential had already been produced. The "container" is his potential for creation, the "handle" is his strength, and the 'flour' was his power, the 'void' – in this metaphor – was his capacity. As I explain in RS1, that potential created the physical on the material level, but the psychic at the atmospheric level. And just as things become conscious within the Mind, what became conscious on the atmos level was a diabolical ego. That is called "evil," because there can be nothing worse than having thought about being separated from the Whole.

The physical universe comes from a spiritual universe, and in between there are other psychic levels, or the Atmos. The souls have come from Ethe through that spiritual universe, as Yeshua says in the Gospel of Mary Magdalene regarding the soul: "I have been precipitated into a world from a world, and into an image from a heavenly image." In that previous universe it was voluntarily decided to come to the worlds that would be created in matter and it was known that in matter there would be a destiny, that is, apart from pleasant things there would be causes and effects, vulnerability, suffering, ignorance, pain, uncertainty, free will, grief, heartache and conflict. In essence, that is the difference between the Shamaim archetype and the Aretz archetype: both are developments within the projection, but the Aretz level is where we perceive suffering.

Returning to the first book of Moses, verse 1 of the first chapter ends with the word 'Aretz', from which come the words referring to 'land' (see RS2, chapter 2 - The Settlers of Sumer, p. 54). The word Aretz derives from the Sumerian 'Eridu' (land founded in the distance), but according to many remote sources it seems to evoke each and every one of the physical worlds that have been produced – and are produced – in creation (the universes). The second line of the Sumerian Enuma-Elish says: «šap-li-iš am-ma-tum šu-ma la zak-rat», referring to that of "below" – which it calls in this case 'am.ma.tum ', that is, Earth - had also not been called by a name, that is, it was not yet structured or defined; and it is important to note that the original Earth is called in Sumerian 'Tiamat' - even though here it is not yet called that, a connotation that reinforces the deduction that it speaks of the primordial mass from which Earth and other "earths" (worlds) were "built" -.

The story of Moses does not begin by telling us about the original void - which is so prevalent among conventional scientific theories - but rather that periodically the universe seems to have been constituted, first from a primeval spiritual universe and, after this, the universe that we now experience. In this way, the question about what happened before the Big Bang would be answered by referring to a root universe from which this cosmos – or dream – and perhaps many others come. It is here, at this point, that Moses would have written the following...

וְהָאָרֶץ הָיְתָה תֹהוּ וָבֹהוּ וְחֹשֶׁךְ עַל־פְּנֵי תְהוֹם וְרוּחַ אֱלֹהִים מְרַחֶפֶת עַל־פְּנֵי הַמָּיִם:

THIS NEXT VERSE TELLS us, " ve.ha.aretz haitah tohu va.bohu ve.choshech al-pnei tehom ve-ruach Elohim merachefet al-pnei ha.maim. " This could be translated into a first section as, " and the earth was in chaos." However, bearing in mind that at no time did he tell us how the Earth was formed - even with its resemblance to the design of the rest of the spheres that extend throughout the universe - we must ask ourselves if 'Aretz' effectively identifies our world or matter itself.. In other words, Aretz can evoke the principle of atomic composition that would have been established in space, or the very scenario for future life that would develop in the cosmos. If so, Aretz is the universe is material composition, but what was that chaos in the story? In theoretical paradigms, chaos is understood as the beginning of nothingness, the unpredictable, the complexity of the supposed causality in the relationship between phenomena or the reason for the fate of the universe.

Chaos or Randomness

The ancient Greeks were already well versed in philosophies about this conception, and in all chaos they seemed to identify the source or force that governs the cosmos in a certain space, under certain parameters that do not follow the model or laws of the rest of the system. In other words, the progression of the development of the universe would be accompanied by the existence of cosmic laws that push energy and order and establish the order of everything. This means that there is a primordial force that in space seems to order all things and give them laws (gravity, electric charges, magnetism, etc.), and through a driving force it causes vortices that condense and repel (like a positive charge). and another negative), keeping everything by force of attraction and repulsion in a spiral shape. That would apply from atoms to galaxies themselves. Chaos would have been the initial part in the "void" where even that energetic force or "intelligence" had not given thrust to the vibration that produced the fields, and it would still be all space in the infinite where that "creation" in progression I wouldn't have arrived.

This version of history is similar to that of many other peoples, only with different nuances, since it follows the same scheme, telling us about that "tohu va.bohu" which are also words with Sanskrit roots that have a sense of "abyss." and "disorder" (absence of a principle of order or law). It makes sense that it says 'Joshej' (darkness), if we accept that in the Big Bang theory, before this massive explosion appeared, there had to be a space where it was produced, or where all the aforementioned potential was projected. Therefore, first there must exist a "place" and then the explosion that occurs in said "place." Consequently, if the "potential" of the "material" universe was not yet there, said pre-existing space was in the dark. Although other texts such as the Hebrew Jubilees, 2nd Baruch, 2nd Ezra or 2nd Enoch vary in one detail or another in the order of events, the pattern is basically the same as that followed by Moses in Genesis 1, and at this point he already speaks of "water" and then "light" (other manuscripts put light first and then water).

Water and Gas

Anyone who does not know the most accepted theory of the formation of matter in the universe also ignores that this approach coincides with the Greek translation of this biblical passage. The word 'water' in Greek is 'idor', from which comes 'hydros', and hence 'hydrogen'. Matter is made up of atoms; Atoms form molecules through valences, and all this follows the law-order principle that we have been talking about. The scientific community considers that the primordial matter is the gas that covers the most in the cosmos: hydrogen. Hydrogen is - according to the conventional scientific proposal - the base substance that gives rise to matter, so it is correct to assume that "the world came from water", as the Bible says.

« ἡ δὲ γῆ ἦν ἀόρατος καὶ ἀκατασκεύαστος καὶ σκότος ἐπάνω τῆς ἀβύσ σου καὶ πνεῦμα θεοῦ ἐπεφέρετο ἐπάνω τοῦ ὕδατος »

The last word of this quote, which corresponds to the Greek version of the same story, is 'Idatos' (water), in the same way 'Idor' and 'Ydros'. Water is twice as much hydrogen as oxygen. Rock crystals, water, ice and most of the gas scattered in space come, according to modern theory, from the atomic principle of hydrogen, since it is the simplest atom, that is, the primordial structure and matrix of the matter in the universe. Being so, by telling us in Genesis that there was a 'Ruach' (wind) 'Merajefet' (fluttering), it seems to indicate clearly that a spiral or vortex movement was moved or pushed in all parts of space to condense matter. That "motor", by the Law of Thermodynamics, had to follow a conscious pattern and a driving force coming from an energy in conservation and intelligent action. In other words, in a strictly classical 19th century scientific-pragmatic sense, vacuum, nothingness or randomness have no consciousness or intelligence, so they could not then create gravity, light, atoms or electromagnetism.

So the consciousness of Mind acting in intelligent infinity produced by vibration the charges of intelligent energy that configured everything from the very basic atom: hydrogen. Following this influence and impulse, the hydrogen would have produced condensation of water, clearly frozen due to the temperature of almost Absolute Zero. The movement would have created friction and friction energy, thus causing these vortices to condense solid matter - largely frozen water -, evidently with the rest of the sums of elements that we have studied in the Periodic Table in Secondary School. However, the vast majority - and the base - would be hydrogen, ice, water...

In the midst of that darkness, what could be seen? Nothing. Without stars in space, what can you see? Nothing at all (unless it is not seen with carnal eyes, then the energy fluctuations would be seen). This is what defines the Greek word 'Aoratos' in the quote above: invisible. The matter or substance - the existent - could not be seen. The Vulgate version of Jerome adds that it was 'Vacua' (empty), and all these concepts coincide with the Hebrew - and even the Aramaic - version, which says « tzadia ve.rokania » (deserted and barren). I had told you that the Enuma-Elish - a Sumerian story preserved in several tablets found in the ruins of Ashurbanipal's library in Nineveh (now preserved in the British Museum), and dating back to 669 BC. or 627 BC. - begins by telling that «When above, Heaven had not yet been named, and below, the Earth had not been mentioned by name, nothing existed except Apsû, the ancient, its creator, and chaos, Tiamat, of the that everything was generated. The waters were agitated in a single group and the pastures had not yet formed and the reed beds did not exist. When no star could yet be seen, none had a name when the destinations had not yet been established.» (Lines 1 to 8).

Ironically the Earth is first a mass, and what became the mold or model was the 'mu.um.mu' (chaos) - at that time Ti-amat -: «mu-al-li-da-at gim -ri-šu-un» ("from whom everything was generated", or "the mother of the two", referring to Apsu and Tiamat). The phrase, "meš-šu-nu iš-te-niš i-ḫi-qu-ú-ma" (waters in a single group), makes one wonder, where and how were they configured? Who are Apsu and Tiamat? According to experts, Apsu is a masculine concept alluding to the "primordial waters" and fresh water, while Tiamat is a feminine idea of the abysses and primeval chaos – the later source of salt water. This would be the equivalent of the Hebrew 'tohu va.bohu'. If we look at the etymological root of these definitions, we find that from Apsu came the Semitic Ab.Zu (wise father), from which later 'Abisu' was derived, and then the Greek 'Abisos', the English 'abyss', or the Spanish 'abyss'. What's more, the southern regions - or the Southern Hemisphere - were called by the Mesopotamians, 'Abzu', as the idea of "what is below". As for Tiamat - or Tiamatu -, it gave rise to the Semitic form 'Teemut', and that to the Hebrew 'Tehemot' (abysses), whose root is precisely the particle 'Tehu' or 'Tohu', from the endiadis 'tohuvabohu' (Interestingly in German, this concept - defined as 'Tohuwabohu' - is a colloquialism for "chaos").

Lightning in Infinity

This idea of "the waters" agitated "in a single whole" is in line with other sources on the same topic of the aqueous mass that was produced from the condensation and vortex fusion of hydrogen with other basic elements that began to compose the baryonic matter. And what was driving all this? That 'Ruach' (wind, spirit), or consciousness-intelligence. At that time, before time, is when the visible appeared - because previously darkness loomed over everything, as there were no stars - as so many sources tell us, including line 7 of the Enuma-Elish: "e-nu-ma ilâni the šu-pu-u ma-na-ma » (when no stars could yet be seen).

וַיֹּאמֶר אֱלֹהִים יְהִי אוֹר_וַיְהִי־אוֹר:

Here Genesis 1 tells us, " ve.iamer Elohim iehi aor va-iehi-aor " referring to the fact that it was light, and it was light. How was that light produced? In line 9 of the Enuma-Elish it also speaks of this: "íb-ba-nu-ú-ma ilâni ki-rib ša-ma-mi" (Then the stars were made visible in the middle of the sky). Although, it is usually believed in creationism that the universe was made in 6 periods (for some this is even literal, or in the best of cases a reference to 6,000 years), but the first period did not begin to be marked but rather from this time, so the above happened before this definite time. In this order of things, only in the 4th period would the stars have appeared, not in the first. However, if the stars were supposed to appear in the 4th period, what was that "light" (Aor) that took place before the clock of time began to tick?

Moses did not contradict himself, and in another work of his authorship - recovered from the caves of Qumran - he tells us that the god, "For the first day he created the heavens and [what] is above the earth and the waters and all the spirits that serve before him [...] and the depths of darkness, [...] (and night), and the light, the dawn and the day, [...] seven great works that He created on the first day." First day? The Hebrew term that is translated as 'day' is 'Yom', but it identifies a period and location where light dominates. The word Yom is associated with the Greek "eon", like Age or Era. The Spanish word "day" derived from the Greek, this being a qualifier for the name Zeus (God), and this reminds us that the ancients related the gods and angels with the stars, just as others translate line 9 of the Enuma Elish: « Then the Gods were created in the middle of heaven. » In the version revealed to the Hebrew scribe Ezra, it relates: « then there was the spirit, and darkness and silence was everywhere, the sound of man's voice was not formed. Then you sent a light just to come out of your treasures, so that your work could appear." (2nd Ezra 6:39-40)

The Big Bang? It seems like a description that both supports the classical version of the origin of the universe and that of other cultures, which maintain that everything arose from "another place" and was provided from there by means of a great light with its spiritual or invisible forces. who started the creation. The Chinese and Hindu version of this story is similar, considering that things came from a kind of "primordial place" that opened and produced everything: the Vedas describe it as an egg that broke and from there came the "deities." " that produced the sky and the earth. It is acceptable to consider that when saying "earth" does not refer explicitly to this planet at first, but to all those in the same conditions, which are not asteroids or comets or stars, and meet certain parameters that include them in the category of "planets." (celestial bodies that follow an orbit). In fact, in the version of the prophet Enoch, the Earth was produced from a thick dark liquid mass that fragmented into 8 parts, leaving the one containing the densest waters configured as our planet Earth, while the other 7 became the others. planets of the solar system.

וַיַּרְא אֱלֹהִים אֶת־הָאוֹר כִּי־טוֹב וַיַּבְדֵּל אֱלֹהִ' ים בֵּין הָאוֹר וּבֵין הַחֹשֶׁךְ

Verse 4 tells us, " Elohim sees et-ha.or ki-tob ve.ibdel Elohim bein ha.or ve.bein ha.joshej, " referring to Elohim seeing the good of this "light" and performing a distinction or separation between light and darkness. What had originally been coming together was now separating; The substances obeyed the force of gravity and the centrifugal force. According to the words of the prophet Enoch, the god of the Hebrews took him off the Earth and showed him how he made the universe, and narrating about this apparent "Big Bang", he tells him: «I ordered [...] that visible things should come down from the invisible ones, and Adoil (light of creation) came down very grand, and I beheld him, and he came down. He had a belly of great light. And I said to him: Be undone, Adoil, and let the visible come outside of you. And he came undone, and a great light came out. And I was in the middle of the great light, and just as light is born there from light, a great age came forward, and showed all creation, which I had intended to create. And I saw that that was good. And I established for myself a throne, and took my seat upon it, and said to the light: Go therefore up on high, and establish thyself high above the throne, and be a foundation for high things." (2nd Enoch 25:1-5)

Enoch seems to describe the story from another perspective, arguing that first there was this light that emanated from who knows where and dissolved or exploded making the invisible seen as visible, but after this God commanded solid matter to arise: "And I summoned the very I came down a second time, and said: Let Arjas (spirit of creation) come out hard, and he emerged hard from the invisible. And Arjas came out, hard, heavy, and very red. And I said: Be open, Arjas, and let it be born from you, and he was undone, an era came forward, very great and very dark, carrying the creation of all low things, and I saw that that was good... » (Ch. 26:1-3) Later Enoch speaks of this 'separation', saying that God told him: «And I commanded that there should be taken from light and darkness, and I said: Let it be dense, and it was so, and I separated it out with light, and it became waters, and I separated it out with darkness, below the light, and then I made the waters firm, that is, the unfathomable (bottomless), and I made foundation of light around the waters, and I created 7 circles from within, and imagined the waters as wet and dry crystal, that is, like glass, and the circumspection of the waters and the other elements, and I showed each of them their path, and the 7 stars each one of them in their Heaven, that they go like this, and I saw that that was good. And I separated between the light and between the darkness, which is to say in the middle of the water here and there, and I said to the light, that it should be day, and to the darkness, that it should be night, and it was late and tomorrow was the First Day. (Ch. 27)

Notably, it is impossible for conventional history to think that anyone would give such descriptions more than 4,500 years ago, so this prophet could not have been making it up. Its description, archaic - it is understandable - due to its time and the limits of its own language, is brief enough to understand that it tells us how after the appearance of that light that spreads throughout the universe - or Big Bang -, matter It condenses in the middle of space and is seen as abysmal masses of hydrogen-based condensation. His description of how light and darkness separate to shape water as such can only be explained as the structuring of water due to the principle of atomic valence of the gases that make up water and also produce electricity. Since these words did not exist in ancient times, it was to be expected that they simply referred to this as the invisible that sustains things or is the essence itself, that is, the Ruach (wind, spirit). That Ruach that "moved" over the water was the law or force that was "curdling" the matter.

In addition to this, the mention that Enoch gives in chapter 27 of his Second Book regarding water and its form suggests that he was told that the invisible and driving force caused the masses to consolidate into "circles" by rotation. As you will see if you review the beginning of this Article, you will see that I already mentioned that the word 'Aretz' seemed to designate the physical worlds created, at least, in this universe. Although Moses only spoke about this quickly, Enoch qualifies it in a formidable way (see RS1, chapter on Genesis 1, page 127), letting us know that the god had already told him thousands and thousands of years ago how the driving force that comes from Him produced the Big Bang, the forces of the universe, the charges that created the atoms and later the molecules and, consequently, the gases from which water was produced (liquid or not, pure or not). Although at first glance it seems that Enoch describes the formation of what could be the planets of our solar system, it is in chapter 30 when he tells us about these 7 space bodies ordered to be fixed in their respective orbits, and what can be interpreted is that The 7 designates a pattern.

By saying that he created 7 spheres one inside another, it leads to the assumption that he is talking about a worldview of 7 heavens or dimensions, and/or a pattern of 7 levels subject to the geometry of the sphere as an elemental model of the universe. This same model would have accommodated the structure of our solar system, although, according to this description - and those that Sumerian sources seem to provide - it would have been different from how it came to be later, having been reorganized thousands of years later to have the distribution that today they own. What's more, when talking about these 7 spheres - and each one in its own sky -, it suggests that there are at least 7 heavens or that there are 7 existential realities in the cosmos; but above all it reminds us of the 7 planets of the solar system in their orbits. Does this mean that you would only talk about the creation of our system? According to these words, God would call this the "heavenly circle", and he states that he ordered that in these "spheres" or "circles" there would be "luminaries", and following this pattern he made everything in "all the heavens ". This manifests a sublime truth: a model or scheme of sequences of 7 in each state of reality with sub-levels also of 7 scales. This is a fractal principle based on the much idealized number 7 that we see in other teachings.

וַיִּקְרָא אֱלֹהִים ׀ לָאוֹר יוֹם וְלַחֹשֶׁךְ קָרָא לָיְלָה וַיְהִי־עֶרֶב וַיְהִי־בֹקֶר יוֹם אֶחָד

Although I already covered many of these explanations in 'RS1' and 'RS2', there is a lot of data that it is always good to review, restructure, update and expand. Here is verse 5, which tells us, « ve.ikrá Elohim la.or Yom ve.la.joshej kara lailah ve.iehi-ereb ve.iehi-boker yom achad. » You just have to see that light is projected in a prism of 7 colors or diffusions, and light being knowledge-awakening in metaphysics, the 7 colors or electromagnetic spectrums come to identify 7 levels of consciousness. As a result, time is divided into 7 cycles, which correspond to 7 states of consciousness, or levels of consciousness. These separate the being from darkness, or in other words, from ignorance-separation-suffering, that is, from death. If I translate the recent Hebrew quote into Spanish, we would find that the designation or identity of the 'Aor' (light) then becomes 'Yom' (day, aeon, era), while that of 'Joshej' (darkness) becomes 'Lailah' (night). Observe that on a metaphysical level between Yom (day, era) and Yam (sea) is experience, and therefore the symbolism of the sea as immensity has intrinsic the idea of time-experience in matter.

At this point, the whole set is known as 'Yom Ajad' ('Day One' or 'Eon Ajad'), and if we remember pattern 7, it is reflected here again, since 6 is spoken of as formation time, and the 7th as the end of this cycle to follow others. Yes, because the creation of the universe did not end in a 7th cycle: it prevails, it continues to take place. The Nag Hammadi texts define these cycles at the highest level, what we could call the "causes", and logically the "effects" would derive from them. There what is referred to is called 'the Battle of the Seven Heavens', which would correspond to part of the 'Lilá' vedanta, or galactic conflagration. Likewise, pure dualism would be projected here, pitting light against darkness. In that cycle of the aeon it would have been defined who was of the Right and who of the Left. Iao, or Jeu (the angel of light), coming from the first 5 aeonic trees, as well as older Melki-Tzedek, and both would be the strong ones who would have maintained control over chaos and its archons. They would be looking over the Middle Region - correspondence with the "between-lives" or intermediate states - while the chests would be on the Left, as an analogy and connection with the lower states of vibration.

After 'The Battle of the Seven Heavens' - say the Nag Hammadi manuscripts - the brothers Jabaoth and Jabraoth separated. Jabaoth is also defined as 'Tzabaot Adamas' (or 'Adamas the Tyrant'), and was the one who ruled half of the 1 2 kingdoms of the aeons of the Heimarmene. According to Ev. Valentine, this tyrant was chained to the region of the sphere by the angel Ieu. Jabraot converted and became a missionary of light, and was known as Tzabaot the good (or simply 'Zeus'), who in other texts is called 'Adonai Tzabaot' (Lord of Hosts). From other manuscripts we understand that Ialdabaoth created 7 sons and 7 potentates, and put each one over 12 aeons, there being some understanding here that Tzabaoth Adamas - the tyrant - is presumably one of the main sons of Ialdabaoth. After the defeat in The Battle of the Seven Heavens, Ialdabaot was thrown into the abyss of the lower planes, where he would remain as rector of the infernal states that the Mind conceives as an inheritance due to its unconscious guilt.

One might think that Tzabaot the Good – possibly linked to the one known as Adonai Yaheveh in the Tanakh – would have been the idea of separation that occurred in Sakla, an analogy of what happened at the beginning, but this time inverted. Expressed another way, this would represent a state of healing from the fall of Pistis Sofia. The Ev. Valentino talks about monsters that Adamas the tyrant creates - including a 7-headed basilisk, something that reminds us of the devil of the Apocalypse -, possibly denoting that the number 7 here is the archetype of the Kundalini serpent. Thus, 7 deep levels of darkness separate the dual mind from the return to the Right Mind. What's more, Vedanta Hinduism identifies 7 higher levels, but also 7 lower levels: the Thala.

Pistis Sofia's deficiency now is healing through the idea of justice that awakens after 'The Fall'. We would be observing that there is a principle of consciousness that becomes aware of its reality and begins the return home, to God. This awakening occurs precisely on the part of the Mind itself, which was apparently divided and which believed it was outside of God. Now the ego found itself in self-doubt. This is a principle of compensation, because where there was a Fall there must be Restitution. From there comes the idea of

Karma, or inertia of actions that are put into operation until they are balanced. Karma stops through Forgiveness, and Forgiveness comes through understanding and awareness of what? That there is no separation, but only oneness, and therefore the Mind seeks the affiliation to which it belongs, even despite the mirages and tricks of the ego. Therefore, Karma and Forgiveness are inseparable concepts. It is an evident parameter of causes and effects, because where the death-separation of the light that conceived individuality appeared, must, then, be compensated as love-forgiveness-light that restores the deficiency.

Therefore, all projections of separation are all results of suffering: old age, illness, ailments, misfortunes and accidents, poverty and death. All of them are compensated by Forgiveness, and therefore healed. Forgiveness also applies through the processes of Understanding, Acceptance and conscious Forgiveness.

Part III

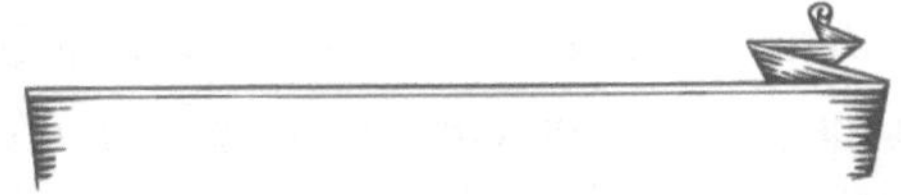

THE FIRMATION OF THE HEAVENS

« Nothing real can be threatened.
Nothing unreal exists.
In this lies the peace of God."
(A Course in Miracles – Introduction, 2:2-4)

THE RELATS

"Before the beginning there was only one consciousness, that of 'the Eternal' whose nature cannot be expressed in words. It was the only Spirit, The Generating Being that cannot be reduced. The Unknown, Unknowable A melancholic solitary in deep pregnant silence. The name that was spoken cannot comprehend this Great Being who, remaining nameless, is the beginning and the end, beyond time, beyond the reach of mortals, and who in our simplicity they call God. He who preceded all existed only in His strange room of uncreated light, which remains ever unquenchable, and can never behold it by any comprehensible eye. The pulsing drafts of light, eternal life in their maintenance had not yet been released. (The Book of Creation, The Kolbrin. The records of Celtic culture).

This majestic explanation of the Kolbrin wisely summarizes for us everything previously said, and expands our vision of this story: "He alone knew, he was without contrast, incapable of manifesting

99

himself in nothingness, since everything within his being was, [he] unexpressed potential. The greater circles of Eternity had not yet come out, to be cast out like the centuries and centuries of the existence of [the] substances. They were to begin with God and return to Him completing in infinite variety and expression. Earth did not yet exist, there was no wind with the sky above them; High mountains did not arise, nor was the great river in their place. Everything was formless, motionless, calm, silent, empty and dark. No name had been named and there were no fates foretold.

The old Greek accounts tell us that "Before there was the great sea and the fertile land, and the blue sky that covers the world; before nature - which our eyes see and all our senses help to accept - lived as it does now: organized, plastic, wise, powerful; Before all that, it was Chaos: rough, shapeless mass that made up the universe. In the beginning, what existed was inert – says Ovid, Latin poet -. It was dead weight. Lots of disparate items. At that time, no light gave the world warmth and clarity. Neither the Sun nor the Moon crossed the celestial vault, transforming each day into a new day, and each night into a clear night. The Earth was not yet suspended in the air, balanced by its own weight. And Amphitrite, the queen of the sea, had not yet extended her sweet arms to the shores. Land and Sea were an indistinct mixture of life and agitation. The ground had no density. The sea did not flow. The air had no light. Nothing had its own shape. And inside that single mass, the constant battle of opposing principles took place: the cold fought the heat; humidity against drought; lightness against weight. Little by little an intelligent germ, a computer god emerged from Chaos. He defined (delimited) and harmonized (balanced) everything, according to his sovereign will. Peace was made in the universe. But the spark of conflict remained lit forever, because order, limits and balance are not static..."

The Quiché Indians also tell us the Mesoamerican version of this story: «Here is the story of how everything was in suspense, everything calm, everything motionless, everything peaceful, everything silent, everything empty, in heaven, on earth. Here is the first story, the first description. There was not a single man, not a single animal, bird, fish, crab, wood, stone, cave, ravine, grass, jungle. Only heaven existed. The face of the earth did not appear; Only the limited sea existed, all the space of the sky. There was nothing gathered together. Everything was invisible, everything was still in the sky. There was nothing built. Only the limited water, only the calm sea, alone, limited. Nothing existed. Only immobility, silence, in the darkness, in the night. Only the Builders, the Shapers, the Dominators, the Mighty Ones of Heaven, the Procreators, the Begetters, were on the water, scattered light. [His symbols] were wrapped in the feathers, the green ones; their [graphic] names were, then, Feathered Serpents. They are great Sages. Such is heaven, [such] are also the Spirits of Heaven; Such are, tell yourself, the names of the gods. Then the Word came; He came here from the Dominators, from the Mighty Ones of Heaven, in the darkness, in the night [...] they united their words, their wisdom. Then they showed themselves, meditated, at the moment of dawn...» (Popol Vuh. Ch. 2)

In their time, the Nordics already said that the world began as a frozen mass called Niflheim, and with it came the world that we can define as that of magmatic activity, which they called Muspelheim, as well as the chaos or abyss that existed between the two: Ginnungagap. Likewise we have the idea of the universe being produced with its God from a golden or shining womb, a cosmic egg called Hiranyagarbha, of which the sacred Bhagavata Purana of India narrates great things. This is how the heavens, worlds, their orbits and dimensions, planes of reality and even parallel spaces and realities would have been produced, and in them the great God, under many forms and representations, who inhabits everything - says Persian literature - "he who he dresses (using as a cloak) the solid stones of Heaven. And He also chose as many as were pleasing to Him" (Yasna XXX, the Avesta – Zoroaster)

The manuscripts of the Egyptian library of Nag Hammadi tell us about these stories in their various treatises, exposing that the "gods" emanated in the chaos, in that darkness and the watery mass. Although, in Coptic and Koine Greek the word for 'gods' was more precisely 'exoisias' (authorities) and 'arjais' (archons, principalities), which later create the 'dynameis' (powers, forces, energies), and with those who represent those powers of Fate. Ironically, although the Hebrew word 'elohim' is translated as "gods" or "God", it really means "powerful", and can evoke "powers" or "forces" behind the movement of the material world.

In another treatise of these Gnostic Christian records, we find: «Sem, since you are of unmixed power and you are the first being on earth, hear and understand what I am going to tell you, [and] for the first time about the great powers that were in existence in the beginning, before the present. There was no light or darkness and there was no Spirit among them. Given their root they fell into oblivion - which was the ungenerated Spirit - of which I reveal to you the truth, about the powers. The Light was believed to be complete with hearing and speech. They were united in one way. And darkness was the ruach (wind, spirit) in the waters. His mind was engulfed in a chaotic fire. And the Spirit among them was a gentle and humble light. These are the three roots. They reigned each in themselves, alone. And they covered each other, each with his power. (Paraphrase of Sem 1:19 to 2:10) All of this agrees with the rest of the versions that deal with this story of the origin of the universe, since the idea of an Eternal and Immeasurable God who existed before the universe and from whom everything comes, It is present in many cultures.

The Vedanta records of India say that "Death was not then, nor was anything immortal. That without breath, breathed by its nature: apart from it, whatever it was, there was nothing. Everything that existed was empty and formless. What was above or below then? Who certainly knows it and can declare it? Where was this creation born and where does it come from? The later gods were the production of the world. Who knows then where existence began? He, the first origin of creation, whether he formed it or not, whose eye controls the world from the empyrean, He certainly knows it, or perhaps not..." (Rig Veda) These people called the powers and stars that emerged "gods" with the "Big Bang".

Day Two

The story of Moses, in verse 6, tells us, "ve.iamer Elohim iehi rakia ba.toj ha.maim ve.iehi mabdil bein maim le.maim", that is, that there was a space or void between the water, understand that it already exists there. He adds that this is done by separating between "waters to waters", that is, waters upon waters. Who hasn't heard the story that God created the universe in 6 days and then rested? Objectively, he does not speak of days but of eras (see RS2, chapter 1, The Ancient of Days, page 34). Once the matter condensed in space, like an apparent unlimited sea, a new cycle of formation began, that of the consolidation of what we would define as aqueous bodies. The masculine and feminine conceptions derive from the essential virtues of the Ethe, and from them the Yin Yang in the dual universe was configured. This is how the Sumerian myth of Tiamat and Apsu (female and male, or salt waters and fresh waters) is understood. We find the same dual masculine-feminine pattern in Genesis codes right at the beginning of the story: remember that we read "barashit bará elohim", as the first three words of the entire Bible. The first letter of 'Barah' is Beit, and the first letter of Elohim is 'Alef', and since in Hebrew there are no vowels and you can read the words from back to front and even swap the letters, it can be read with these two initials 'Ab' (father), likewise, the last letter of Bará is 'Alef', and of Elohim is 'Mem', which form 'Am' (mother).

וַיֹּאמֶר אֱלֹהִים יְהִי רָקִיעַ בְּתוֹךְ הַמָּיִם וִיהִי מַבְדִּיל בֵּין מַיִם לָמָיִם׃

The para-biblical Hebrew text of 2nd Esdras (chap. 6:41) tells us: «On the second day you made the spirit of the firmament, and you ordered it to be divided in two, and then to make a division between the waters, which on one side could rise, and the other will remain at the bottom.» This description is clearly an allusion to the formation of the atmosphere, at least in our world and those that are similar spheres. How would this occur? It is a logical action that responds to the law of gravity: some elements are heavier than others, and the lighter ones (gases) overlap the heavier ones (liquids), and these even heavier ones (solids). Jubilees 2:4 confirms this fact, arguing, from the words of Moses: «And on the second day he created the Rakia in the midst of the waters, and the waters were divided on that day; half of them passed over and the [other] half of them went down below the Rakia (which was) in the middle over the entire face of the earth. And this was the only work [that] (God) [had] created on the second day." (Jubilees)

The Children of the Dawn

The ancient 'Emerald Tables' of Thoth tell us about "the Lords of Amenti, lords of the Children of the Morning, Suns of the cycles, Masters of Wisdom." (Table III) Dyehuty (Thoth), the Egyptian "god" of writing, stated that these lords of Amenti controlled chaos so that the universe would not be dissolved, and, according to him, the existence of chaos was necessary to maintain a balance in the cosmos. Dyehuty said that these immortals of Amenti (the rooms of the path to the afterlife) were lords over the "children of the morning" and in turn "suns of the cycles." The Idumean wise man Job wrote, some time after Abraham, "when all the morning stars praised together and all the children of God rejoiced..." (Job 38:7, KJV 60) Did the stars praise? In Hebrew it says 'Kokabei Boker' (morning stars), the same ones that Dyehuty seems to mention, and of whom he added, "Many of the stars I passed on my journey, many of the races of men in their worlds; some reaching high as the morning stars, some falling low in the blackness of the night. And he also wrote that "Many ages ago, the Morning suns descending, found the world filled with night, there in that past, the fight began, the ancient Battle of Darkness and Light." (Table VI)

Now, these "suns of the cycles" identify the control of time, since the sun represents a period of time – the Mayans themselves called their eras "suns." The Hebrew word Yom (day, era, eon, age) refers to an era and a god, as encompasses the Greek word 'Aionos' (eon, era, century, age), also the Hebrew 'Laila' (night) has its own deep meaning. Its Sanskrit root, 'Lilac', identifies it as the concept relating to the eternal struggle in the cosmos, since darkness emerged. For the Vedas, the Lilac was the eternal cosmic battle between good and evil, the fight between order and chaos, light and darkness, justice and injustice. The very name of the demon Lilit - from Jewish legends - takes its name from the same root, being the idea of evil, darkness and the demonic. Another important point here is that the narrative of Genesis 1:5 tells us of a 'Boker' (morning) and an 'Ereb' (evening) that make up that first aeon or initial era, but, if the sun is not supposed to have yet I was, what were they that morning and that afternoon? Furthermore, it does not say "night", but rather the day would only be according to the period of light, the principle of 6 hours in the morning and 6 hours in the afternoon? Consequently, it does not speak of a 24-hour day, but of a light cycle. If I were talking about a consecutive period of actual days, I would have had to include the night. This agrees with the other sources that state that this light was not only cosmic (from spheres of light), but from beings of light that govern the positive side of that Yin-Yang, or "Lilá", compared to the chaotic ones that govern the dark side.

We can mention here that those "stars" were extraterrestrial and interdimensional beings; angels from superior spheres of reality that fight against darkness and its forces, and those "suns" or "sons of god" are also a high degree of supernatural intelligence that operates on order in the cosmos. In the "dawn" periods of civilization they were here, some of them having been deserters in pre-Flood times. Genesis tells us that the Yom is divided into two sections: Boker and Ereb. Boker's are maximum light and from that era, but what is this 'Ereb' thing? The twilight of that era (cycles that began and ended with decisive episodes), of which the ancient Greeks recorded when they said that in the remote era of the world, 'Erebus' was thrown into hell, where he became the horde of demonic spirits that in that remote past devastated the world.

The Rakia

Enoch's version is even clearer on this story, telling us: "And then I made the heavenly circle firm, and caused the low waters which are under Heaven to collect themselves together, into a hole, and that the chaos became dry, and it became so. Out of the waves (wave) I created hard and large rock, and from the rock I filled the dry, and the dry I called earth, and the middle of the earth I called the abyss, which means the bottomless (unfathomable)., I collected the sea in one place and limited it together with a yoke. And I said to the sea: Contemplate, I give you your eternal limits, and you must not break, let go of your component parts. Thus I quickly made the firmament. And there was evening and morning on the Second Day. (2nd Enoch 28:1-4) The "heavenly circle" that was formed when the parts of lighter water (gases) rose from the rest that remained below, was that void (Rakia) between both, and the rest of Matter from the bottom covered what was originally a hole.

This is what is also told in the Enuma-Elish, where the story is described as a pitched battle between gods to refer to the way in which the solar system was formed. In verse 5 he says: " meš-šu-nu iš-te-niš i-ḫi-qu-ú-ma", which translated would be: "the waters were agitated in a single group". After this the gods, or the "heavenly beings" appeared: " Then the Gods were created in the midst of heaven, and Lahmu Lahamu were called into existence... Age increased,... Then Ansar and Kisar were created, and upon them..... Long were the days, then he did not go out..... Anu, his son,... Ansar and Anu... And the god Anu... Nudimmud, to whom his fathers, his begetters..... Abounding in all wisdom,..." According to the long description given by the Sumerian tablets, various impacts of space bodies and formation processes led to the establishment of the solar system as we know it, but the primordial mass Lahmu and Lahamu changed state and "spirits" emerged and subjugated.

Precisely the words Lahmu and Lahamu give rise to the Hebrew word Lechem (bread), due to the idea of a pasty dough that is evidently then heated to obtain the precious food. The earth as a quagmire and also fiery (still hot and with many volcanoes and lava) began to produce its atmosphere in a long cycle of changes with the processes of consequent change, pressures and revolutions of the terrestrial sphere. That idea that Enoch describes - of chaos becoming dry or solid - agrees with the other passages of the story of Moses, so we can argue that the changes followed one after another, but the sequence was not necessarily according to the end of an era to start the next one, but even many starting before finishing the other. How does this whole story differ from the scientific version of the geological formation of the Earth? Actually practically nothing, except that these people already knew it thousands of years before current science.

Biblical translators translated 'Rakia' in various ways, from 'vault' (Reina Valera of 1995) or 'expansion' (Reina Valera of 1960), but from the translation of the Seventy the idea of 'Steréoma' was accepted (firmament, firmness, firm), as Jerome did much later in Latin: 'Firmamentum' (Vulgate version). There are many texts, whether Hebrew or not - especially from the prophet Enoch - that address the issue of how the "spirit" of God formed the atmosphere - or "heaven" according to "biblical" vocabulary - and designed it with this perfect spherical silhouette covering the surface of our world. The word itself, Rakia, derives from the root 'Rek' (void), since in a way the layers of the atmosphere, although filled with gas, appear to be a vacuum above us, as if we were inside a large transparent balloon. filled with air or helium, because the gas is invisible, even though it takes up space. Now, all that gaseous mass did not cease to be part of those primordial waters, even though at first glance our perception of things does not make us imagine this as such. They are just compositions to a greater or lesser extent of the same elements.

This situation even responds to the hydrological cycle, where the magical clouds play a transcendental role, and that invisible "spirit" that motivates life carries the wind currents from the south and the north, from the east and the west, whether in the atmosphere or in the ocean floors; It maintains the clouds that were born with the atmospheric formation at its height thanks to the pressure and the climate, it promotes precipitation that nourishes the land, it takes the fresh water down to the sea and again the sublimation in the ocean takes the particles back. liquid towards the clouds. The text rightly says that "let the waters separate from the waters."

וַיַּעַשׂ אֱלֹהִים אֶת־הָרָקִיעַ וַיַּבְדֵּל בֵּין הַמַּיִם אֲשֶׁר מִתַּחַת לָרָקִיעַ
וּבֵין הַמַּיִם אֲשֶׁר מֵעַל לָרָקִיעַ וַיְהִי־כֵן׃

IN THE SEVENTH VERSE of Genesis 1, the Hebrew text tells us, « ve.ias Elohim et-ha.rakia ve.ibdel bein ha.maim asher mi-tachat le.rakia ve.bein ha.maim asher meal la.rakia ve.iehi-ken", which in a few words means that this "firmament" was made and the "waters" that were under the firmament were separated, which remained that way, but also the "waters" above, on that " empty". It can be assumed that certain "waters" ended up beyond this world, into heaven, as some of us have once suggested, giving formation to other planets (as is the case of the "aqueous" or mega "gaseous" ones that are more beyond the Earth: Jupiter, Saturn, Uranus and Neptune (a planet that, however, is identified as "the sea of heaven" in Vedic culture, and that the Greeks called 'Poseidon')). However, what can be considered the easiest is that we are talking about O3 (the ozone layer). Just that layer? No, the context of the atmosphere as a gaseous mass. Remember that there is always talk of levels, so an interpretation of one level does not refute the one that corresponds to another level.

So that, just as before Day One (first era defined, Aeon One) was the spiritual universe, later the "void" in this universe, then the appearance of light (Big Bang, stars, etc.) and later the matter being configured, in the Second Aeon the "Aretz" or physical masses were structured, which thanks to the law of gravity took the spherical shape and separated its substance by weight, giving rise to the outer planetary layers: atmosphere. Moreover, in the work 'The Twelve Planet' (The Twelve Planet) by Zecharia Sitchin, he determines that the gods of the Enuma Elish (Anshar, Kishar, Tiamat, Apsu, Lahmu, Lahamu, Nudimmud and Kingu) are also references to the planets of the solar system, along with the Moon (whom Sitchin associates with Kingu). Not all planets seem to follow the same pattern and be a paradise, but the idea suggests that they follow the same pattern (which according to the laws of physics would make a lot of sense).

וַיִּקְרָא אֱלֹהִים לָרָקִיעַ שָׁמָיִם וַיְהִי־עֶרֶב וַיְהִי־בֹקֶר יוֹם שֵׁנִי׃

Verse 8 tells us, " ve.ikra Elohim la.rakia shamaim ve.iehi-ereb ve.iehi-boker yom sheni ", which defines that the 'Rakia' received the title, name, designation or identity of 'Shamaim'. This is how this Second Aeon concluded, but the problem here is that before the "days" began, it had already been said that the Shamaim - as well as the Aretz - had been created. What's more, explanations were given about the state of that matter, or 'Aretz', that came after the Shamaim, so this "other" Shamaim seems to emulate the first one or be an "inferior" image of it. If you observed in the previous quotes, there was constant talk of the "name" that was before all things and from which they emanate. Precisely "name" in Semitic languages is 'Shem', which is also read as 'Sham' (over there), and which gives rise to the word 'Shamaim'. Although in Hebrew "names" is 'Shemot', in the feminine plural, not in the masculine plural, it is worth noting that 'Shamaim' - or 'Shemim' - was an idea of 'names' (identities, destinies) projected from that original Shem, original Shem being said the engine of all the others.

Many manuscripts tell us that the sky is full of skies, and what do those skies consist of? Obviously this is a precarious definition that encompasses what we now know how to define the universe with many words. If this Shamaim is the entire space that encompasses the atmosphere of the planets, their orbits, their dimensions and/ or all of the above, these descriptions make fascinating sense, since they fit perfectly with all the scriptures that address this matter. Even the descriptions about "vortexes" or "spirals" explain that movement of the "spirit" flowing in everything and motoring everything: "As you see the power of the whirlwind collecting the dust of the earth, and driving it together, know that in the same way the ji'ay, a'ji and nebulae are gathered in the firmament of the heavens; by the power of the whirlwind I can create the corporeal suns, moons and stars. And I told the man to name the whirlwinds in the Etherean firmament, and he named them according to their shape, which he qualifies as vortices and wark." (Oahspe, 1882 revelation)

You can study this concept of the heavens with the data provided by Vedanta, and it gives you detailed information about the 7 lower levels - or Thalas - and the 7 upper levels - or Loka - of this planet. I will focus on the 7 levels of our consciousness, since understanding the levels is very useful on the path of awakening consciousness. In this sense, if one transcended in an astral, mental journey or with a new garment-body of glory, one would first surpass this Rakia (or firmament), and from there one would cross the first sphere towards the higher realities - the circle of the spheres. - which is the expansion of the powers of these systems. It would then rise to the door of the second sphere, which is the Heimarmene - or region of destiny -, and above it is the region of the 12 eons of these planes. This is the region of the Great Master of heaven and the so-called three great Triple Powers. Logically, all this is under the veil, in line with the lower realities, but at the highest levels already under the veil. In that region is the first Adamic tyrant, lord of the powers that have ruled there as a personification of the ego in said interstellar region and dimensional state.

From the beginning when these powers wanted to lord it over the creation manifest in the Cosmos, the angel of light (Iao, Ieu) gave them all a limit and location within these regions, to create order, as Moses said, " and separated the light from the darkness." Upon reaching the region of the lower aeons (those under the imperishable realm), there are the veils of the 13th aeon, or 13th region of the aeons. These aeons of the powers that emanated from the power of Pistis have here their state of highest dimensional level, and here resides the last and greatest of the rebels, behind which is the veil to the higher realities. These levels, worlds, heavens or Lokas, are higher vibrational states elsewhere in this galaxy. Jewish culture defines them as the 6th and 7th heaven in the classical version, which was also believed by some Catholic mystics. However, thanks to the prophet Enoch, we can consider that the idea of the 10 sefirot has an equivalence with 10 heavens, where the last - or Arabot - would be the state and region above the veil. In this way, said 13th eon was considered to be the 9th heaven, the region of the 12 eons is the 7th, and the region of the Heimarmene is the 6th.

This description gives us a certain connection with the various stories about heaven, expressed in numerous epigraphic texts of Judaism. All analogies about evolution are applicable to the level system, and the most basic example of all of them is that of the tree, like Yggdrasil in Norse mythology. It is also known as the tree of life, and in its case it would be a perennial ash tree whose roots connect the 9 worlds, which is an analogy of the 9 diffusions of consciousness that Dyehuty mentions in The Emerald Tables of Thoth. Even so, these concepts of the heavens are nothing other than references to the levels of consciousness, and the Rakia of the Shamaim is nothing other than the veils that exist in the Mind between the unconscious, subconscious and conscious aspects. In this sense lies the elementary principle of why there are hundreds of thousands of religious denominations worldwide, since the texts on which they are based are not that many. Simple: they apply what they read at the level of form, images and projections, not at the level of Mind. This is what the apostle Paul was referring to when he said that he was still wearing the veil that the Jews had. The same veil that still prevails in Judaism - to a large extent - is the one that passed into Christianity and largely prevails, except in certain minorities, such as certain students of ACIM, or those who correctly recognize the roots of Gnosticism, to give an example. couple of examples.

Understanding the Scriptures

"There is nothing hidden that should not be known nor secret that should not come to light," means that the unconscious (secret) manifests itself in the conscious (the light). The unconscious is the programs of the mind, and the conscious is the results or manifest aspects. It is the same as saying "every tree that does not bear fruit is uprooted and thrown out", where the tree identifies knowledge, experience and the mind itself, the good fruit are the results of the Right Mind; The uprooted is the undoing process carried out by the Holy Spirit; what is cast out is the end. I will show you some cases where you can see the notable difference between the truth revealed by the Holy Spirit about the Mind, and what you yourself will be able to recognize as quotes that have been interpreted arbitrarily by Judaism or Christianity – if you are familiar with such beliefs or have you ever heard of any of these concepts.

Let's see, when one reads the history of biblical Israel, one finds that less than 4,000 years ago the patriarch Abraham had 13 great-grandchildren (one girl and 12 boys), whose father was the Hebrew patriarch or Yakob (or Jacob). Yakob's daughter was named 'Dinah', which means 'judgment', and I was raped by a prince of Sikem (from the Hebrew 'Shacham', meaning shoulder) named Hamor, which means "donkey". It is usually spoken of the 3 patriarchs: Abraham, Ytzchak (Isaac) and Yakob. This is symbolized by the parents of the incorruptible kingdom, which are Esefec, Adama and Set. After them came 12 kingdoms, which are represented in the 12 tribes of Israel, since Israel is the name used to identify the Sethite race of the imperishable kingdoms upon arrival in the world. Who identifies, then, Dinah? She is the idea of the Greek Athena, produced by Zeus (God) from his own mind, but the mistaken projection of the separation is the Ephaestus son of Rhea, the wife of Zeus. That is to say, the Dinah who was raped, since the rape was nothing other than having been outraged by the ego – or call it in the Christian religion, devil.

All sacred texts reflect the levels of Mind over illusion, so any part of the Bible that you wish to interpret in the light of the Right Mind – which is the Holy Spirit – will explain the truth to you, whereas if you wish to understand it at level of the form-illusion, it will only lead to assumptions and new theological tendencies subject to dualism. Now, it is said that Israel was 12 tribes, that is, the Setite race in 12 imperishable kingdoms, but 10 tribes were dispersed by the Assyrians. Once again we see an illustration – at the level of the form – of something that happened in the Mind. In this case, it is

Shalmaneser of Assyria who personifies the ego, and the 10 tribes identify the Setites entering the illusion and projecting themselves in the dream in various parts of the cosmos (since Jewish theology maintains that the 10 tribes are no longer in Nineveh, but they were mixed by the nations). I had already told you that the Hebrew word for nations, people and Gentiles is one and the same: Goim. Goim derives from 'Gei', that is, relative to the Earth. Since Aretz (earth) identifies the material cosmos, or physical planes of the first dimensions of this octave, ergo, the majority of Setites were withdrawn into worlds of the first levels of density.

But what about the tribes of Juhdah (Judah) and Benjamin? Much is said about Judah, but practically nothing is said about Benjamin. In Hebrew, Ihudah means "praise Iah", and is, in fact, a code for IHVH where the letter Dalet is inserted in the penultimate position. It is or means that through such 'Ihudah' the access door of the god is opened. But it is not the Jewish people he is referring to. Normally one speaks of 9 and a half tribes, since the Levites were distributed among the 11 tribes. In fact, Levites, Jews and Benjaminites symbolize 3 groups, and the rest - called Israel or lost tribes - is a 4th group. The 4th group are those who are furthest along the path back to oneness. That is, the beings of the universe who do not yet understand the role and integral part of oneness.

dream projections. The Benjaminites, as their name indicates, are those who remain anonymous, but are serving the truth: because Benjamin is 'Ben-Iamin', or "son of the right", which means "those of justice". Then there are the true Levites, who fulfill a priesthood knowing that they are not part of the world. These 4 groups are the Setites in the stages of development where they are at a 3rd density level, some performing 4th density services. For this reason, the yehudim are usually defined as rebels, since they are the same as the lukewarm Christians of the New Testament, or people who are aware that they are part of the truth, but do not finish defining themselves. In this way, the Levites and Benjaminites are the "righteous", while the "chosen" are taken from the ihudim (Jews) - which Christians have incorrectly assumed refers to them. The chosen ones are not of blood or flesh, nor of religion, but are those beings aware of the truth who decide to live according to the truth. Both they and those who are lost in the darkness of the ego - called Goim - recognize at certain moments in their lives their superior origin.

The difference between goyim (gentiles) and Israelites is that the goyym have come from a much lower level of consciousness development, while the Setites called Israel came later, but aware of their role. In theory, the so-called Israelites or Church, or congregation – manage to awaken before those metaphorically called Goim. But I am speaking from the level of Mind to the level of the psyche, not of the people that the Jews call goim and Israel. Therefore, the so-called "rebellion of Israel" because of their "idolatry" and "fornication" is actually the manifestation - at the level of form - of the works called "sin", which are developed by the consciences within the dream. Fornication is the consequences of listening to the ego and how it "warms your ear," as the Spanish say: they put ideas in your head. This leads to guilt, which activates karma due to subconscious and unconscious guilt. For its part, the so-called idolatry is idealization and was placed in the figures and illusions of the world. There are many references to this in ACIM, but I will refer to this one:

« Well, by standing in front of this idol and seeing it exactly as it is, you make a choice. Are you going to restore to love what you have tried to take from it and place it at the feet of that inanimate block of stone? Or are you going to invent another idol to replace him? For the god of cruelty takes many forms. It is always possible to find another one. Ergo, don't think that fear is the way to escape fear. Let us remember what has been highlighted in the text regarding the obstacles that peace has to overcome. Of these, the last, the most difficult to believe that is actually nothing, although it appears to be a solid, impenetrable, fearsome and insurmountable block, is the fear of God Himself. 4Here is the basic premise that enthrones the thought of fear like a god. 5For fear is revered by those who worship it, and love now seems to be clothed with cruelty. (A Course in Miracles, Lesson 170, VIII and IX).

The primary concept that Yeshua explains in ACIM refers to anything in the world, since everything is images, forms, projections and phenomena that derive from the Mind, and therefore, are delusional creations of it. All of them are the projection of illusion, while God is the real, what represents the spirit. For this reason, section XII later in the course refers to the fact that by correcting your Mind, opting for the Holy Spirit instead of the ego: "You have chosen Him instead of idols." Therefore, idolatry is the belief in the world and its mirages, and since an idol has no life, it cannot provide anything real either. In the same way, anything in the world is just an image of our Mind - not something real - and consequently believing in an idol (projection) is disbelieving in the truth, because you think that there is something external that exists, that has life that has some kind of power. Worse, the world is not real, therefore it has no power, no life or action over you: you are the one who creates what happens in your world, not the other way around.

When spiritual Israel was still in God, it was united, but upon listening to the ego, there was rebellion, and therefore they were dispersed into the cosmos that was produced. Thus, such "punishments of Iaheveh" are nothing other than the effects of the causes derived from the unconscious guilt of the Mind for feeling separated from God. There is no real god punishing anyone, because god is not dual, he is not insane. The only god who is punishing is ourselves. It is the ego who says that Yehovah "will destroy his enemies," without knowing that that phrase comes from the Holy

Spirit to refer to the ego's own thoughts: those will be the ones that will disappear, and "will cease to be forever." The Holy Spirit will progressively eliminate all our thoughts of separation, because to be separated is to be opposed to the light of truth, and the opposition is called in the Hebrew language 'Ha.Satan'. That is the accuser: the ego. It's so ridiculous that it makes us feel bad for the projections he himself has instilled in us. The ego itself told Yeshua "get out of here", as if Yeshua believed in the dream; and told him "turn the stone into bread", Yeshua being fully aware that the stone is not real, nor is hunger, but the truth. Only the truth, which is the One, is real.

Then the prophet says about Yaheveh, that this god would have referred that "for love of my people I will repent", since the Holy Spirit refers to the fact that love causes forgiveness to heal illusory projections. It does not refer to a special love but to a holy love, which is understanding the oneness of which we are all a part. We are the ones who, when we understand love, change our attitude, because we understand that there is no one to blame, since we are all innocent, as the Course (UCDM) says, because "we are all the holy son of God." In other words, all of us are the only begotten, since God only has one son. That is, we come from God, since we are him, and he is in us. We are God, and in the Hebrew language "son" is not only the descendant, but the integration of a group. Therefore, being a "son of god" means that one is a god, part of the divine lineage. When we understand this, we stop pointing and condemning, understanding that phrases like "the sword of judgment" are nothing more than the experiences of karma that heal and balance us. So anyone who hurts us is not just a reflection of our own unconscious flaws, but a portion of our being that is not yet aware of what has happened, and acts only on the impulse of the programs within its unconscious, adapted to your belief system.

The Tanak says that "by the fulfillment of this law you will prolong all your days on earth", which means that the application or lifestyle according to the truth of love and oneness will make you receive the experience of eternity, our part in the imperishable kingdoms – from the place from which we apparently detached ourselves. Yeshua represents that truth, for he was given the knowledge of the mysteries of the aeons and the production of the universe. That is why the apostle said that Yeshua will save the remnant of Israel - referring in the spirit - to the fact that Yeshua is in charge of the atonement, that is, of the plan for the healing of the Mind, and therefore with the Holy Spirit he is in charge of those who They must awaken and understand the oneness.

Another example from the Bible and the Mind is when it says that "he will gather them from the ends of the earth", meaning that the Holy Spirit works in the work of the return to oneness, the sonship back to the Father. A similarity of this work of Yeshua is seen in the prophet Ihoshea (Hosea), who also bears his name, as Ihoshua (since that is how Yeshua was called from birth). Ieshua is the same as Yeshua, as Yerushalaim as Yerushalaim, or Ihudeah as Yhudeah. But look how curious it is that the prophet Ihoshea (Hosea) was sent to marry a prostitute, why? So that they would relate the prophet to Iaheveh, and the prostitute to Israel. That is the same as saying that Yeshua married a prostitute, which is what is being integrated into the oneness. But first comes getting to know each other, then the fashion ceremony and, finally, the union in the bridal chamber. In the bridal chamber they become one flesh, as Pistis (Eve) must have been with Christ in the eternal mansions. Another Biblical example of the same thing – because it is ALWAYS the same, one thing and only thing – is the log that was thrown into the water and turned the waters sweet.

The classical understanding of Christianity only idealizes things to worship Yeshua more, perpetuating with its attitude the intrinsic aspect of the uniqueness of the message. Precisely Christianity only makes the understanding of a certain level of these symbols a legitimization of duality. Because they do nothing more than guide it towards separation, as do the rest of the religions. Both wood and wood are the same in Hebrew, and their analogous symbol is the cross and the tree, but it is not so evident that these ideas identify Mind and its levels. The tree is not only used to represent a genealogy, but to symbolize the paths of the transcendence of being. For this reason the tree or wood is - along with the monolith or stone - a primary element of idolatry. This means that the focus that one places on a tree identifies its focalization, and therefore, the orientation of its life. Logically you understand that I am not talking about a physical tree, but about the archetype of the tree. However, Yeshua, Buddha, Shankara, Krishna, Lao Tse, Confucius, Nichinen Daisonin, and others like them, are models to follow, and figures that have contributed a system of teachings based on uniqueness, filiation, love, and forgiveness., with more than less excellent guidelines, but above all, useful. That is the wood, the knowledge of the truth, since each tree has its "own properties", but all identify a deep knowledge.

Therein lies the phrase "my people failed for lack of knowledge", referring to humanity and the knowledge of the truth. And the prophet adds, "because you have rejected knowledge, I will reject you from the priesthood", which means that the Holy Spirit cannot add members to the filiation if they do not seek the truth, and those who have it in their hearts, but they do not apply it and do not make it real – ceasing to be mere theory – they become fruitless and useless to be light. And it says elsewhere that "they, just like Adam, trespassed my pact", which means that the thought of separation remains in force in those who do not wish to wake up. On the contrary, whoever practices true forgiveness and true love, and thus "teaches smaller ones" is in a state of greater work in the awakening of consciousness. Those "little ones" are not of size or age, but what come just a couple of steps back on the same path of awakening.

This is how the phrase of the Tanak that refers to "the soul of the priest will be satisfied with wine" is applied, since wine is the joy of knowing and understanding the truth that everything is just a dream. The priest symbolizes the servant who is aware that he wants to wake up from the dream and, meanwhile, help other brothers to understand the projection and the path of uniqueness.

For this reason it was said that "the king comes on a donkey's colt", since a king is the Malkut state, or lowest state of consciousness of being. The one who from Malkut ascends towards the truth through humility (symbol of the donkey, which in turn heals the separation represented by Hamor de Sikem by violating Dinah), as expressed by Yeshua himself upon entering the city – symbol of the state of oneness – riding on a donkey. We must follow all of Yeshua's guidelines, since he is the model, since "son of man" is an identification given to all of us who are the only begotten son. Yeshua does not represent, therefore, each human being, in what I want to express, but rather the entire group, the entire son of man, the entire Adam-Adamah. Another passage maintains that "never again will there be two kingdoms," as we will all return to oneness. That is, there will be no more duality. And it is also written that "you will live in the land that I gave to your fathers", which are the imperishable habitations (for those fathers are the Geradama). Elsewhere he maintains that there are privileges for "those who love God," that is, those who live a spiritual life. What does it mean to live a "spiritual" life? The transfer of the experience without losing sight that everything is a dream, and that we actually seek to have our mind in the state of awakening, of silence, of union, of love and forgiveness, since the opposite is " flesh", that is, of the world (part of the illusory dream).

We can see the same when reading that it says: "seek first the kingdom of God and his justice". All the needs within the projection come to us alone from reality when the first thing we do in our life is to seek truth and uniqueness, which is understood in our own balanced being and in being one with our peers, helping them to awaken to be in the same vibration of consciousness. As the apostle says, "it is better to give than to receive", because whoever wants to receive does not receive if he does not give first, and what is giving but serving? He who first serves, then receives, for we reap what we sow. It is the same as the passage that refers: "commit your way to the Lord and trust in him," and then adds that "he will grant you the requests of your heart." Do you trust him about what? That you seek your inner awakening and that of your peers and he takes care of your personal affairs that are part of the dream script. This is what Yeshua was referring to when he said "do not worry from morning to evening about what you will eat or what you will wear," or also when he stated, "work not for the food that perishes, but for that which endures to eternal life."

One more case, that of the prophet from whom it was translated: "in two days he will not give life, on the third day he will raise us up and we will live before Him." RVA does not translate "will rise" but "will rise", but in reality the two concepts are the same, and refers to the awakening of consciousness, illumination and the growth of the inner being. The two days in which it gives us life are the two cycles of primordial development of the being before its experience in the third density. The first day is the experience of being in inert and cellular forms, and the second in vegetation and animals. However, it is in the third cycle where we acquire self-consciousness, and therefore reach transcendence and Nirvana. That is why the texts say that "the son of man was resurrected on the third day", since in fact man will achieve awakening in this state of third density, and will move to the fourth density. This is what identifies the constant appearance of the concept of the number 3 in biblical numerology, and the explanation of another "half-time" in which the transition is reached (since it is not possible to simply begin the 3rd dimensional state).

One of my favorites is Paul's classic, "for Satan himself masquerades as an angel of light," identifying the cunning wiles of the ego to elude us, making us consciously unable to differentiate when we are listening to the ego or to the Holy Spirit. It is the classic question that comes to our mind on numerous occasions: how do I know that this comes from my Right Mind (from the Holy Spirit) or from my Wrong Mind (my ego)? The ego's reasoning tends to be persuasive, to the degree that we can be convinced that a supposedly good intention is the right thing to do, when in reality it is a ploy by the ego. Another example is when the Israelite scripture says: "sanctify your hearts." In ancient Hebrew culture the word 'Leb' (heart) was used to refer to the source of emotions, as in demotic (ancient Egyptian), with the Ib. although, in demotic this idea was more related to the justice and integrity of being, coinciding with the fact that Leb is, more deeply, the representation of the ego as a source of ideas and thoughts. Ergo, sanctifying the heart is the same as correcting the Mind consecrating it to the Holy Spirit, making no concessions regarding the absolute truth that nothing in the world is real.

Finally, to give one last example and move on to the next chapter, let's look at that famous eschatological prophecy of Gog-Magog, where it says that their bodies will be buried over 7 months... the same as before. Gog represents the ego, and just as Yeshua was "tempted" by his ego at the end of his last trial, so it is with the Collective Mind within the dream, where "fire falls from the sky and consumes" all wrong thoughts. As a result, the light of truth will descend on ignorance and the idea of separation and the ego will disappear. But the Mind is very powerful, and its fixed beliefs are not erased by magic, so it takes 7 states of dream experience to completely erase the distortions of separation. Then the chosen ones will rejoice and birds will come and eat the bodies of their enemies, since other members of the sonship work from other dimensions to eliminate the ignorance and blindness of our minds. The "body" of the enemy is every structural misconception of dualism. Thus, the elect "will rejoice and drink and eat," that is, correct thoughts will produce prosperity, joy and other consequences of sonship and uniqueness. Now phrases like: "and everyone will know that because of their rebellion they were scattered" make sense, do you get it? Do you see the part of Sofia and the Mind that broke away. And "we will know how we were known", since we have already known ourselves since the end of history, but I will explain this to you later.

Part IV

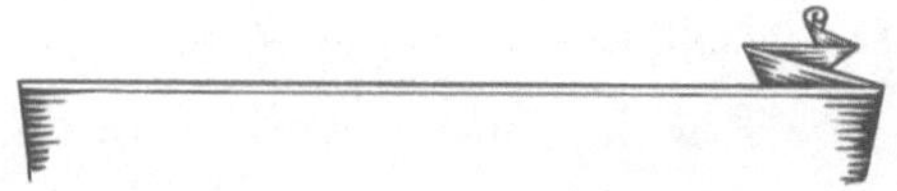

A WORLD OF ARCHETYPES

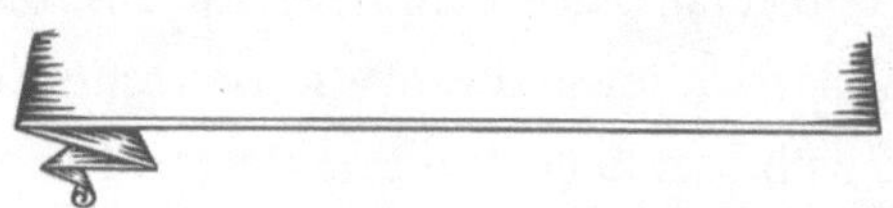

«Since the stars fell from the sky
and our highest symbols faded,
governs the secret life of the unconscious."
(Carl Jung - Archetypes and Collective Unconscious 1)

Day Three
We pass from the state without time and without space to the state without time in space; later there was consciousness in space, focusing to produce intelligent energy toroids; consciousness became conscious by atomizing and molecularizing. There was an explosion of light and this dream began. Another version would invert these two facts, but if we did not invert it it would seem that before the Big Bang like a white hole, on the other side an initial matter was produced that was absorbed by an amazing black hole that swallowed that darkness and released it as transmuted energy to This space. Well, this is just speculation, although some astrophysics theorists have postulated it. In any case, matter was formed from gas, or vice versa - depending on who wants to propose it, because in any case that is not the point, but rather the metaphysical aspect -, and the spatial bodies were created, and then the compositions were configured. planetary. So we come to the part where worlds received the first glimpses of life and their geographic and environmental parameters were formed.

When the atmosphere was forming, land also began to appear. Enoch's version told us that at the end of the Second Period, just as the gas cover of the Earth was formed, the rock mass that would build Pangea emerged from the depths. While this was happening - according to the version of this prophet - the angelic beings were created, and one of them rose up and dominated the abysses: " And for all the armies of heaven I imagined the image and essence of fire, and my eye I looked towards the very hard, solid rock, and from the flash of my eye the lightning received its wonderful nature, which are both fire in water and water in fire, and one does not put out the other, nor does one dry up the other, for Therefore lightning is brighter than the sun, softer than water and firmer than hard rock. And from the rock I cut a mighty fire, and from the fire I created the orders of the armies of tens of thousands of uncarnal angels, and their weapons are fire and their armor a burning flame, and I commanded that each one must keep in his order. And one of the chief custodians of the angels, was obstinate (twisted) with the custody downwards and promoted [an] impossible plan: to erect his throne above the Earth considering his strength comparing it with mine, and the consequences of his conceited with his angels, and he was [and] prosperous unfortunately on the face of the Abyss, always." (2nd Enoch 29:1-4)

וַיֹּאמֶר אֱלֹהִים יִקָּווּ הַמַּיִם מִתַּחַת הַשָּׁמַיִם אֶל־מָקוֹם אֶחָד וְתֵרָאֶה הַיַּבָּשָׁה וַיְהִי־כֵן

Moses' version in Barashit (Genesis) 1:9 says: " ve.iamer Elohim ivakú ha.maim mitajat ha.shamaim al-makom ajad ve.teraeh ha.ibashah ve.iehi-ken ", which refers to gather the aquatic or liquid mass that was under the sky (that is, what was under the atmosphere) in a single place, and the dry part was visible. In other words, the most solid part was consolidated into dry land, but how, if water covered everything? In 2nd Ezra 6:42 it tells us that "on the third day, you commanded that the waters be gathered into the seventh part of the earth: six parts you dried up, and maintained, with the intention that some of these [would be] Hewn [and] were planted by God and could serve you. The objective - according to the text - was for that 7th part of the world to be solid land to be cultivated, and consequently, for the purpose of serving as a home for humanity. But, if humanity did not yet exist (at least on this particular globe) it is necessary to assume that humanity was destined to come to worlds where the habitats coincided with the level of vibration of the being, that is, its state of density., its dimension.

The idea that the earth makes up only 7th of the earth's mass would not have been accepted by science if it had not been for the discovery of underground oceans that was made a few years ago. Thanks to this discovery, it was learned that this planet is not strictly a mass of rock covered in water, but rather that this distribution is not precisely uniform, possessing even more water on lower scales of the crust. Jubilees 2:5-7 - a text also attributed to Moses and found in the caves of Qumran - maintains: "And on the third day he commanded the waters to pass from the face of the entire earth in one place, and the dry land appeared.". And the waters did as he commanded them, and receded from the face of the earth in a place outside this Rakia, and dry land appeared." The question of how this would have occurred naturally is part of the enigma of plate tectonics. Although it is really a theory, what several scientific proposals propose is that the force of gravity and the change in pressure and temperature of the elements of the lithosphere with respect to the asthenosphere cause geological changes and continental displacement. Although, the radiation and heat of a star would alter the gases and underground pressure causing the movement of masses.

A CHARACTERISTIC GARDEN

If this position is correct, it coincides with the previously mentioned approaches, since vortex forces, the force of gravity and changes in temperature and pressure raise the light parts and sink the heavy ones in a permanent cycle. This is how mountains and continents have been created - as in the case of this planet -, and this process would have given birth to Pangea. While the rock was pushed up by these forces, the water moved sideways, in the direction of the space it was in (because two things cannot occupy the same space at the same time), giving "birth" to the ocean (maritime mass). only one that has received the scientific name of Panthalassa). According to accepted chronology, all this was the end of the Paleozoic period. But the end of Yom Shlishí (Third Day) had not yet arrived, rather, beginning the Mesozoic era, life "emerged": «And thus I made all the heavens; and it was the Third Day," says Enoch when referring to his entire previous string and the creation of the beings of those worlds and their dimensions, and adds that "on the Third Day I ordered so that the Adamáh would abound with large and prosperous trees and the mountains, all of sweet grass and all semina (seed) on what was scattered; and I put a garden and closed it, and guarded [with] nurturing-attentive angels of fire.» (2nd Enoch 30:1)

Jubilees also speaks of this, adding that "that day has created for them all the seas according to their different meeting places, and all the rivers, and the gatherings of the waters in the mountains and on all the earth, and all the lakes, and all the dew of the earth, and the seed that is sown, germination and all things, and fruit trees, and the trees of wood, and the garden of Eden, in Eden and all. These four great works that God created on the third day." (Chap. 2:7-8) Both writings tell us that when the mainland emerged, plant life appeared in all its extension - or so it is implied -, including the Garden of Eden, but in a progressive period that could last a lot of time. But, if the terrestrial surface was still united, and the continents began to separate, where was that garden? To the east? Taking studies on Pangea as a reference, the "east" was the Tethys Sea (or 'Tethys'), unless it spoke of Eurasia to the north or India and Australia to the south. This suggests whether it is perhaps coinciding with the Oahspe story, which states that the first great continent - called there 'Waga' - was dissolved, and from this specific portion of land where this particular garden was located, what we now call 'Japan'.

Others place the garden in Mesopotamia, where it seems to coincide when reviewing the rivers of the story and connecting it with the Sumerian texts. Although, there was not only one garden at that time, but there was one that seemed to be most relevant, in the image of the kingdom of light where it all began (before the Big Bang). The Spanish word 'Garden' - which is replaced by several Bible translators as a synonym for 'Orchard' - comes from the Hebrew word 'Yarden', which is the name of the river that came to sound in Spanish as 'Jordan'. Observe the symbolism and metaphors regarding this in the Bible, where it is said that half the tribe of Manasseh was divided to the east of the Jordan, then its waters were divided to allow Israel to pass, and on the other side the other portion of Manasseh remained. Who is Manasseh? The brother of Ephraim, sons of Yosef and the Egyptian Asenath. Egypt symbolizes the world of separation, where idealized concepts, or "idols" and gods reign. We live in that Egypt, although our heritage is from heavenly Israel, and that understanding of the two states is the reason for the union of patriarchs with Egyptians.

We know the analogy of Yosef in Egypt with Yeshua, but in itself, this brother who comes just before Benjamin is the "borrowed" or "substitute", the one who is temporarily in a place for a mission that must be completed, and after whom the real one will come. That is why Yeshua's stepfather was also called Yosef. Precisely Manasseh means 'to forget', while the name of his brother Ephraim means 'fruitful' or 'profitable'. These two are the derivatives of Yosef, which once again symbolizes one of the 12 kingdoms. Yosef symbolizes the temporary state of separation, what was broken off from the source, just as he was sold by his brothers. However, where this terrible act came from, salvation came for many, since Adam has experienced the dream that is perfecting him in his own self-knowledge, and when he achieves his fullness he will value what he really is and the source of life- light.

Forgetfulness - or Manasseh - is humanity in the state that goes from the beginning of its projection in sleep until its overcoming of the 3rd density state. The man referred to as Manasseh does not remember who he is until a certain advanced stage of Third Density, when he begins to be aware of who he is. He is divided (on both sides of the Yarden), but when he becomes one again he will be in paradise. The Yarden – from which the English 'Garden' was derived - is represented as a river for the idea of the flow of blessings, flourishing and prosperity, and therefore Yochanan baptized in the Yarden, where Yeshua was also immersed. By immersing themselves there, the participants were being programmed with the idea of re-entering eternal life, but since they cannot "stay permanently" underwater, they come out again, since man is not yet ready to transcend the power of water., that is, the world's dream levels and forgetfulness. For water as matter is also an analogy of the multiple levels and layers of the complexities of the Mind, which are projected in the cosmos as scales of consciousness-learning.

The apostle refers that " Jesus manifested [his glory in] Jordan. The fullness of the kingdom of heaven, which [preexisted] the All, was born there again. He who before [had been] anointed was anointed again. He who had been redeemed redeemed in his turn." (Ev. Felipe 81) This indicates that everything emanated from the garden of life, which they call Paradise (from the Persian 'Pardes'), since it was a beautiful and perfect place, the most beautiful and beautiful, as metaphorically described in Genesis. It was in that paradisiacal garden where the Christ was first manifested by means of his glory, that is, where he first showed who he was to the realities of the imperishable kingdoms. This is what Yeshua did as an analogy, and from there he was known. As the wise Valentino refers in the second century AD, " We have been brought from those on the right, that is, into immortality, which is the Jordan." (Commentary 'A' on Baptism, Valentino).

The Offspring

However, there is another enigma to discover here, and that is, where did the tree seeds come from? We know that there are plants and trees that strictly depend on human care, and that seeds emerge from somewhere, because it is an elementary principle of bio-genesis. It's the same story of, "which came first, the chicken or the egg?" Logically, they were both at the same time (because both ideas were created together in the Mind before being manifested in the dream projection), but in this case, where did the trees, bushes, plants, tubers, mushrooms and other herbs come from? This is where the theory of Panspermia comes in, first formulated by the Swede Svante August Arrhenius (Nobel Prize winner in chemistry), and defended by Hermann Richter - although it was the respected and renowned British astronomer Sir Fred Hoyle who made it popular.

According to the classical theory of Panspermia, life came to this world from space, possibly coming from one or more asteroids that had spores or frozen cells that came into contact with an optimal environment on our globe - a terrain and conditions propitious - produced the later existing forms. This would be like a less conventional theory of Evolution, except for the problem that, as Bio-Genesis points out, life is produced in a similar way that brings it into existence, so even if life had come from " outside" that does not explain the variety of species, intelligence or their level of development, unless a Noah's ark of plants and animals had fallen from the sky. In addition to this, life, although "life is sought", to "survive" - redundancies apply -, depends on adequate conditions to occur, and theoretically bacteria would not survive the very high temperatures and forces involved in an impact. against the Earth.

It is believed that extremophile bacteria would overcome this barrier, and by recreating this scenario with organic molecules - such as amino acids - it has been found that not only are they not destroyed, but they begin to form peptides. The most coherent position is the one defended even by Ancient Astronaut theorists, who, supported by ancient testimonies and records, maintain that life was deliberately brought to our world from other ends, especially from Orion planetary systems. This is how the theory of Directed Panspermia makes great sense and can satisfactorily explain the dilemma of the origin of the cell and, perhaps, that of complex organisms. 2 Ezra 6:43-44 adds: "For as soon as your word spread the work was done. Because immediately large and innumerable fruits were produced, and many sweet and pleasant to the palate, and flowers of unalterable color and wonderful-smelling odors, and this was done on the third day."

The Sumerian version of this story also coincides, implying that the "gods" or beings of light in the universe saw that this orb was already ready to support a paradisiacal garden for the gods. It is curious the relationship that is found when looking at the Mesopotamian definitions for these concepts, such as the contrast between "steppe" ('Edin' in Sumerian; 'Serum' in Akkadian) and "irrigated land" ('Gan' in Sumerian; 'Eqlum' in Akkadian). Although, in Hebrew, 'Gan Eden' means 'garden of Eden' or 'garden of Eden', the Sumerian form 'Edin' means - etymologically - "plain" or "flat place beyond the cultivated lands". It is equally curious that the Indo-European word with the same consonants of the Hebrew GaN, are 'GheN', or simply 'Ghn', which probably derives from the same stem from which the Sumerian form Gan (irrigated land) or the Hebrew (garden, orchard). The Sanskrit phoneme Ghen precisely means "chaos", which later passed to the idea of "disorder", but came to Greek as 'Gen' (Earth), and its variants: Gi, Ge, Gin, Gea, Gis, Gei, Gii.

For its part, the word Eden passed into Hebrew as a colloquialism for "paradise", which in turn derives from the Greek 'Paradeisos', and this from the Persian 'Pardes'. It is even conceivable that the root itself had the same origin that gave rise to the Hebrew form Gai, or Gia, meaning "valley." Clearly we have to understand the same Aretz was first a paradise, but it became a chaos, since it has been the state of creation of the erroneous Mene. All this leads us to the logical conclusion that our sphere was a water world that stood out as a paradise when the mountains and mainland emerged, and exuberant plant life soon filled everything, and apparently especially towards the east. It should be added that the idea of springs, rivers, lakes or seas - that is, "waters" - is referred to in the manuscripts that exemplify the states of glory in which the emanations of the Perfect One are found. These glorious regions are accompanied by these representations, which could explain why Poseidon was interested in creating a glorious kingdom in the middle of the Atlantic Ocean in the image of this.

וַיִּקְרָא אֱלֹהִים ׀ לַיַּבָּשָׁה אֶרֶץ וּלְמִקְוֵה הַמַּיִם קָרָא יַמִּים וַיַּרְא אֱלֹהִים כִּי־טֽוֹב

In this verse 10 the text says, " ve.ikra Elohim la.iabeshah Aretz ve.la.mikuh ha.maim kará yamim ve.ira Elohim ki-tob ", referring to the dry expanse of the world receiving the name 'Aretz ' (Earth, 'Eridu' in Sumerian), while the area where all the waters were gathered into a single mass, was called 'Iamim' (Seas). Oh! Only on the third day would the Earth be named, so once again we see a structure of levels: there was a divine Aretz, a universe of Aretz (material) and finally aquatic planetary spheres from which solid land emerged with its breathable oxygen and plants. for food. According to Sumerian history, Eridu – the Sumerian Aretz – was the first city to be established on Earth, the place that was initially founded when the surface was found to be stable and habitable. The book ' Mesopotamia and the Ancient Middle East' (1992) evokes the words relating to an ancient Sumerian inscription, which says: «Not a reed had grown; a tree had not been created; a house had not been built; a city had not been made; and the lands were sea, when Eridu was created.

Several times there is this description that the world was mostly water, and the first part that stood out was this plain, which seemed to stand out from the area of Mesopotamia to Japan; later the rest of Pangea was coming out "afloat". The legends about the Akpallu or Anedoti, of the god Oannes and other Babylonian narratives, also speak of the aquatic world, of the era of the reign of the sea god Ea (later ruler on dry land as Enki), who built his first settlement in the sea., analogy from the Greek story of how Poseidon built Atlantis. In fact, Enki would have been responsible for the creation of the Mesopotamian rivers, through a long process of drainage, water diversion and canals, similar to the Egyptian myth of Ptah, creator of the Nile (some believe that Enki, Poseidon and Ptah were the same being). But if now the Aretz (mainland) and the Yamim (the total oceanic extension) were clearly defined, with those names, how is it that the beginning of Genesis comes to say that it was then that the "Aretz" was created? And not three "aeons" later? There were some Shamaim and an Aretz that existed – and still exist – before the existence of this universe, and the Heavens and the Earth that now exist are images of those that were before time and space.

In what proportion could this happen? How many times could this have happened or will this pattern be repeating itself in this or more universes? An eternal heaven and earth create a heaven (dimensions, planes, orbits) and an earth (matter) in the universe and those heavens and earth create many like ours: planets (Aretz with their Rakia (Rakiot)). The oldest Sumerian tablet found delights us with these words: «The reptiles truly descended. The earth is resplendent like a well-watered garden. At that time Enki and Eridu did not appear. Daylight did not shine, moonlight did not emerge. The planets weren't even there yet, but the Earth had already begun to blossom and it was a beautiful, beautiful garden. Without sunlight, how did plants grow? How did they photosynthesise? Without light, how was this process of changing light energy into chemical energy to produce carbon carried out? By geothermal energy absorption? By synthesis of inorganic matter? By radiation derived from cosmic rays?

None of this is entirely clear, but there are many records of plants and trees growing in conditions of almost permanent absence of light, or at least to a large extent. They could grow and enter a dormant phase, but in order for growth to occur they need to follow a light source. The Sumerian text maintains that the earth was "shining," but appears to use this epithet in the sense of describing the beauty of the plant habitat. One hypothesis, although somewhat daring, would be to consider that the possible sowers of plant life had brought seeds capable of absorbing the energy of geothermal energy or cosmic radiation, or have introduced seeds created with this capacity. All this sounds a bit crazy, but one factor that stands out about plants and trees lacking light is that their growth increases more than normal due to a normal reaction whose purpose lies in the need to extend their trunk in search of light. If there is one thing that has been instilled in us since we were children in school, it is that the Mesozoic era stood out for the outbreak of life forms of a supernatural size, whether in plants and trees or animals.

The carbon balance on our planet comes mostly from photosynthesis carried out in the aquatic environment by algae, cyanobacteria, red bacteria, purple bacteria, and green sulfur bacteria. This could explain why conditions were favorable before the appearance of terrestrial plants, even when there was no sunlight. The ancient Egyptian text of Hermes Trismegistus, 'Corpus

Hermeticum', tells us that "there is a place, beyond Heaven, a place without stars and separated from all corporeal things. There is another Dispenser that is between Heaven and Earth, which we call Jupiter. As for the land and the sea, they are under the dominion of Jupiter Plutonium who nourishes mortal living beings and those that produce fruit. "It is the energies of all of them that provide subsistence to the land, the fruits and the trees." (Treaty of Hermes to Asclepius. Verse 27) Could these bodies be just being formed and their energy already influencing our world?

I believe that the earth was visited by cultivators and scientists from other worlds who were probably in a more advanced state of consciousness, having passed through our experience many eons ago. The most important thing is to understand the mental part of all this, and in that order of things we can see in that Sumerian passage about the reptiles that descended before anything, an analogy with other similar myths, which are nothing else - at the level of the Mind – that personification of the worst and most sinister unconscious thoughts of separation. Yes, they appeared when the Mind gave rise to an ego, when the Mene was at peace, "like a well-watered garden," because in that then-state it lacked nothing at any level – except for the experience itself of understanding what it is. to be outside the uniqueness with all its consequences -.

וַיֹּאמֶר אֱלֹהִים תַּדְשֵׁא הָאָרֶץ דֶּשֶׁא עֵשֶׂב מַזְרִיעַ זֶרַע עֵץ פְּרִי עֹשֶׂה
פְּרִי לְמִינוֹ אֲשֶׁר זַרְעוֹ־בוֹ עַל־הָאָרֶץ וַיְהִי־כֵן׃

GENESIS 1:11 ADDS THE account of the production of trees, seeds, and fruit as something to be given in a sense of "greenness": "make the Earth green." We can read: "ve.iamer Elohim tadeshe ha.aretz deshe esheb mazria zera etz pri oseh pri lemino asher zaró-bo al-ha.aretz ve-iehi-ken", where there are several words that

are translated differently, although they are the same or hardly varies its root. For example, it is stated that 'Tadeshe' should be made - from the form 'Deshe' (to produce, swarm) - from 'Esheb' (grass, grass, vegetation). As we can see, the cognate of 'Deshe' and 'Esheb' is similar, and something similar happens with the form 'Mazria' (with seed, seed) and 'Zera' (seed, offspring), which accompany these words in this phrase. He affirms that these seeds are from "fruit trees" which in turn also produce products that also have seeds, and that all of these are of different species, depending on the type of seed, and all this occurs on Earth. What does this mean at the level of Mind? Thoughts that do not remain mere ideas, but manifest themselves, and not arbitrarily and sporadically, disappearing later, but causing consequences, learning cycles, karma.

וַתּוֹצֵא הָאָרֶץ דֶּשֶׁא עֵשֶׂב מַזְרִיעַ זֶרַע לְמִינֵהוּ וְעֵץ עֹשֶׂה־פְּרִי אֲשֶׁר זַרְעוֹ־בוֹ לְמִינֵהוּ וַיַּרְא אֱלֹהִים כִּי־טוֹב:

Verse 12 of Genesis 1 tells us: " ve.totze ha.aretz deshe esheb mazrea zera laminehu ve.etz oséh-pri asher zaró-bo leminahu ve.irá Elohim ki-tob ", and we can see that first an order is given, then the order is executed, which suggests a determination that comes from "above" and that is subsequently carried out. That is, it is not something that produces the "mind of God", since, playing to this idea, why would god have to say anything, or give an order to the Earth as if the Earth could create something by itself? This means that the order was deliberately given so that in our world there would be fruit trees with their genetic information that would produce fruits that, according to their own typology, would produce more fruit trees of their corresponding species. Also, if it is something that comes from God, why does he have to "see" that it turns out to be "good" in the end? And is it that perhaps something bad was going to come from God or go wrong? What he implies is that the project that he orders to be unleashed goes well, turns out to be a success. At the level of Mind, she created all things in the cosmos at all levels of consciousness: Unconsciously, subconsciously, and consciously, depending on the level of consciousness.

וַיְהִי־עֶרֶב וַיְהִי־בֹקֶר יוֹם שְׁלִישִׁי

Thus, with these words, Moses ends the story of the things that occurred in that Third Eon, which for us could indicate all the periods of the Paleozoic and Mesozoic era. This is not only assumed by the chronologies of the origin of Pangea and Panthalassa, but by the same fact that the fossils of the most remote trees have been found from periods ranging from 140 to 300 million years. Precisely the end of the Paleozoic was the Permian, about 286 million years ago, and the Mesozoic eras (Cretaceous, Jurassic and Triassic) spanned from 144 to 248 million years ago. That does not mean that the plants appeared immediately when the earth emerged, so whatever time it took for "life" to be planted here and these "experiments" to be carried out successfully, it could have been throughout everything. the Mesozoic. Thus life would have evolved, not alone, but through improvements made by those who responded to the guidelines of "elohim", that is, what the Mind produced through its thinking.

ETHEREAL WORLDS

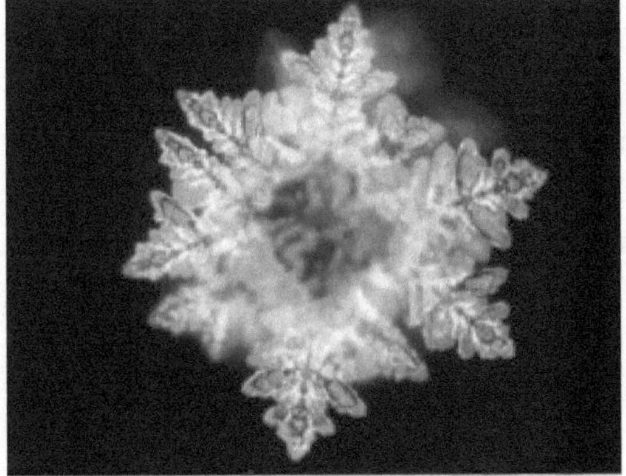

To talk about the Fourth Yom we must understand how perception develops within the dream of this universe. We define that there is a veil between the octaves of this "reality" and the imperishable world outside the dream, and that veil is called "spirit." Since the concept of "spirit" or "spiritual" is overestimated, idealized, converted into ambiguity or taken out of context by ignorance and religiosity, I will define it as they wisely call it in the Oahspe: Es. The very set of letters here agrees with the acronym 'Holy Spirit' in Spanish, which is fabulous and holy (because nothing is a coincidence, and the ES teaches the truth through multiple mysteries and codes). What's more, given that everything is a script already written outside of time, it is not surprising that there are so many synchronicities - as Carl Jung called them - even in the words we have adopted or in the names with which we have baptized objects or things, places or people.

Well, beneath the veil is the physical perception - or 'Corpor' - of corporeal or material forms. In this way, in essence, there are two levels or basic principles of projection: Es and Corpor. 'Es' has two sub-levels, which are Ethera, or Eterea, and Atmosferea, or what I prefer to call 'Atmos', which would be similar to the "world of ideas" or mental plane of the spirit. This Atmos plane has 3 sub-levels *per se,* which are called Ji'ay, A'ji and Nebulae, or Nebulae. What is on Earth is also there, only those there are more rarefied (although the First Heaven of the Earth is strictly Atmos substance). Those who dwell or experience perception in the 'Es' states are usually called es'eans, spirits or angels, while those of Corpor are simply known as "corporeal". This is what the psalmist refers to when he states that they "execute their logos by obeying the voice of their precept" (Tehillim 103:20). This means that in this state they are in the highest level of consciousness, which are the fields vibrations of 4th Density of consciousness onwards.

This Ethe state and Atmos levels is where the consciousnesses are at the level of service to those who are still polarized at levels 2, 3 and 4 of consciousness. Not that all consciousnesses of 4th and 5th density are balanced, but they are the states from which one works to help awaken the other soul projections. Because of that, Psalm 104:4 says: "he makes his malach his ruaj, mashar of flaming fire", or "he makes his malach his ruaj, who are their leaders of flame of fire". A Mashar is a minister, and here the concept of Ruaj is the same as that of a spiritual creature or es'ean. The addition that defines them as ministers of 'Esh Lahet' - or flaming fire - evokes their role as light workers whose function lies in the awakening of consciousness and instruction through revelations and knowledge. The ji'ay, A'ji and Nebula essences gather by random consciousness in the form of whirlwinds, that is, by the toroidal energy of focalization of the Elohim Consciousness, in such a way that they compress the elements producing the physical suns, moons and stars.

The corporeal worlds have been created round, with earth and water, and impenetrable, so that their inhabitants dwell on their surface. However, the ethereal worlds are not round or impenetrable. These worlds resemble snowflakes – hexagonal shapes – and with fractal endings. Their appearance is thus similar to crystal paradises with crystal paths, rainbow colors, arches, curves, angles and millions of angels in meshed hierarchies. According to the Oahspe, it could take a man a million years to travel just one of the Ethe worlds. When those who disincarnate go to the Atmos planes, or lower heavens of their sphere, the structuring of the mental parameters that define their next step in their evolution of consciousness develops, and if they have reached the point of growth of higher vibration they go through the Sinvat bridge from his upper scale of Atmos in the Duat to travel to the Second Heaven, that is, to Nirvania (the Ethe worlds). The Atmos of the Earth (width of the First Heaven) is 1,504,000 miles, with the planetary sphere floating in the center of it.

Just as the most solid or "dense" elements are compressed towards the interior of the focus - creating the so-called gravity - the lighter elements - or Atmos - that configure the First Heaven of each sphere or spatial body. Thus, the suns (also called 'photospheres') come to be corporeal worlds with the Atmos densities than Ethe, and they influence the corporeal worlds and their consciousness level borders. That is, the photospheres are bodily suns as they travel through Etherea (which others call 'Nirvania') and its worlds, since that is their purpose. For their part, the spheres or plateaus that rotate and move with the Earth are called Atmosferea, or 'inferior heavens'. On the contrary, the idea of the Second Heaven is already the remote extension of Towsang (our solar system), so that in reality each planet of our solar system has its own "First Heaven" where its corresponding "lower heavens" are located, 'Duat'. The dividing line between the First Terrestrial Heaven and the Etherea-Nirvania plane, where the Atmos state passes into Ethe is called the 'Chinvat bridge' or 'Sinvat bridge'.

Now, the idea of the recognition of truth, or awareness, manifested first among the gods, thus serving as a bridge to those who would come later - within the illusion of time and space -. As an analogy to this, 5 lights were structured in contrast to the powers polarized towards the SAS. These 5 lights follow the pentagonal pattern - the concept of the 5 senses or mathematics of the π (Pi) pattern - as the fingers of each hand and each foot make evident in our body. This is understandable as an example of heaven, since 5 is the Hebrew letter 'He', and as the 4 He that contain the sphere of the "4 angles of the Earth" and those that contain the "4 angles of heaven", our hands and feet project this into the phenomenal, or sub-lunar, state. The positioning of these 5 space bodies came as a result of the end of the Battle of the Seven Heavens, which began with the formation of the solar system - or Towsang - and defined the aspects of the minor cycles of time perception. So when the Moon and the Sun arrive, a set of 7 is created, once again. This is another example of the system of levels of progression of consciousness, since the Sun is the Greater Light, at the top of the ladder, and the Moon is the Lesser Light, being at the bottom of the ladder. In between would be Mercury, Mars, Venus, Jupiter and Saturn.

Day Four

Since Maldek is not mentioned in the list of planets of our original system, it is understood that it was destroyed before the set and its order were completed, and/or simply, not being a permanent structural part of the group, it would not be counted., because that way it would not be an archetypal symbol to configure the 7-level ladder. The same can be said for Uranus, Neptune and the other outer planets of the system (so-called 'trans-Neptunians'), such as Pluto, Nix, Hydra, Cerberus, Ixion, Haumea, Namaka, Orcus, Makemake or Huya. In any case, it must be known that Maldek was destroyed more than 500,000 years ago, so it would make sense that it had not been included in the list of the cosmology of our solar system given to the prophet Enoch. Now, once these three previous eras have elapsed - since the origin of the universe of which these stories tell us - the moment seems to have arrived in which the great luminous spheres and planets began to fix themselves in specific orbits.

This is how Enoch describes this development: "And on the fourth day I ordered that the great luminaries should be in the circles of the Heavens." (2nd Enoch 30:3) He speaks to us in Hebrew about 'Jugei', from the form 'Jug' (circle), but what are the "circles" of the heavens? Could it refer to the "circular" things in the heavens? If so, he could be talking about planets and stars. For contemporary astronomy, it could be considered that the "circles of the sky" could be stellar mechanisms, such as orbital movements, local trips of star systems, galactic trips of local systems, galactic rotation itself or the movement of local galaxies. It is notorious, since everything in the cosmos carries out rotations at all levels. In the case of Enoch, he seems to describe the same thing as Moses, speaking of the "meorot ha.gdolim", or "great luminaries." However, why does it say "great luminaries" and not "stars" (kokabim)? By kojabim - or Kokabim - both planets and stars can be understood; but the Meorot identify something beyond this, starting with the fact that several puns are implied here.

The word 'Meorot' is a plural of 'Meoré', which in turn derives from the form 'Aor' or 'Or' (light), which means that they are "light bearers" or "light givers".. It would be more accurate to translate it as "those that provide light" or "those that are luminous." Within the phonetic concepts themselves, Meoré resembles the form 'Moréh' (teacher, professor), indicating a similarity with the idea of "guide". 2nd Ezra 6:45-46 tells us that "on the fourth day you ordered that the sun should shine and the moon should shine, the stars should be in order. And you gave them a burden for the service to man, which was to be done. Although it had already been clear with the fruits that the vegetation was going to produce, now the stars complemented this work. But if the analogy between stars and angels was already understood to revolve around yom One and Two, would there also be a relationship here? We could think that the Earth – or "lands" – was provided with rulers to ensure what had to be carried out, and this coincides with the references that we have observed in the processes that took place throughout the Third Yom.

Likewise, it seems to say directly that both the Sun, the Moon and the "stars" were "fixed" by a law. These two texts do not tell us that these bodies were created, but rather that they were ordered to do their "job": to illuminate. This suggests that these stars existed previously but had not begun their work. In other words, both the kojab and the malaak (angels) already existed at this point in the story, but they were given directives regarding the city, guidance and administration of the lower level consciousnesses.

וַיֹּאמֶר אֱלֹהִים יְהִי מְאֹרֹת בִּרְקִיעַ הַשָּׁמַיִם לְהַבְדִּיל בֵּין הַיּוֹם וּבֵין הַלַּיְלָה וְהָיוּ לְאֹתֹת וּלְמוֹעֲדִים וּלְיָמִים וְשָׁנִים:

MOSES WROTE IN GENESIS 1:14: " ve.iamer Elohim iehi meorot ba.rakia ha.shamaim lehabdil bein ha.laila ve.haiu leotot ve.lemoadim ve.leiamim ve.shanim ", and if we understand that time and space revolve around the existence of masses, this passage hones in perfectly, as it maintains that it was said that there must be Meorot in the firmament - or space - of "the" heavens – speaking in plural – with the purpose of "separating" the Laila (night) and being "signs" for the "yamim" (days, cycles, eras, aeons) and "shanim" (years). The positions of these bodies would "divide" the darkness and define the cycles of time. But why does it say "rakia ha.shamaim" (firmament of the heavens)? If there is an upper heaven and another defined as our atmosphere, what is the Rakiah of those shamaim? Definitely the space - or "void" - that exists in all the heavens, which reinforces the already accepted idea that the sky is full of star systems, and re-emphasizes that the heavens are the places that encompass infinity.

וְהָיוּ לִמְאוֹרֹת בִּרְקִיעַ הַשָּׁמַיִם לְהָאִיר עַל־הָאָרֶץ וַיְהִי־כֵן:

In the story of Moses, all celestial bodies seem to be included as 'Meorot', which is why the 'Shemesh' (sun), the 'Iareja' (Moon) and the 'Kojabim' (stars, planets) come into the description. as well as the Mazalot and other Tzba ha.Shamaim. Mazalot is usually translated from 2 Kings 23:5 as zodiac, but Tzba is host or army. Now, is this pattern repeated throughout the universe or only in our solar system? Being methodical in the analysis, we must consider that the name 'Sol' is Norse, while 'Shemesh' derives from the Akkadian 'Shamash', while in Sumer it was known as 'Utu' (the Greeks called it 'Helios', and were not very wrong, since the sun supposedly remains incandescent due to the fusion of helium). The word 'Shamash' or 'Shemesh' means "identifies fire" or "who has fire as his name"; Even Enoch says that it is also called 'Ur-Hamah' ('burning light' or 'hot light'). For its part, the Moon is called 'Iareja' or 'Iareaj' due to the definition of the "month", while according to Enoch it also has the name 'Lebaná' (white); It is called Iareaj for defining the 30-day cycles that separate the year into 12 parts.

Then the Sun sets the theoretical 12 hours of the day of the Earth's equator and the time of a total day (24 hours), the Moon with its phases the stages of the month, while the other stars define the other cycles: the planets of our system mark the more detailed eras and stages of longer, more encrypted cycles, while the star groups we call 'constellations' define the others, and furthermore, the even longer eras and the total journey of the sun, or the complete solar cycle. The cycle of translation in which the Earth was set took its number from the kingdom of Adama, from the god who is Elohim. This parameter - within the Heimarmene - was determined based on the 36 dekan (decans), which govern in 3 the 12 locations of the zodiacal signs, which in turn determine the 12 annual phases that are combined with the Moon. In the Valentinian Exposition the following is commented: "But the Decade of Word and Life brought forth decads in order that the Pleroma became a hundred, and the Dodecad of man and the Church gave birth and made the Triacontad in order to than three hundred and sixty and become the

Pleroma of the year." The 'Three Hundred and Sixty' is the name of the upper aeons of Adamas, as Valentinus also comments, regarding the abodes of the immortals and the imperishable man: "while he dwelt in the Three Hundred and Sixty."

However, the stars define time, but they also illuminate the Earth. 2 Enoch 30:4 says: «In the circle, the first, the highest I placed the star Shabetai, in the second above I placed Noga, in the third Maedim, Shemesh fourth, Tzedek fifth, Kojab sixth, seventh Iareaj; and in the stars, the small ones, the splendid ones, those are the aerial ones, the last-final ones.» These are the Hebrew names of the known planets in our solar system up to 1780, when Uranus was discovered, and then Neptune in 1846. Shabetai is Saturn; Noga is Venus; Maedim is Mars; Shemesh is Sun; Tzedek is Jupiter; Kojab is a star and, likewise, the name given to Mercury; Iareaj is Moon. This order that Henoc gives is strange, since it does not correspond to the sequence of sizes of these bodies or their location in the solar system, so it could refer to their initial state. What is evident is that if we previously asked ourselves about those "circles" of the heavens, these 7 space bodies were distributed from largest to smallest in that order.

In addition to the fact that nothing is said about Uranus and Neptune - at least speaking of that era of the creation of the solar system - mention is made of other "stars" or "planets" that are "small" - logically in relation to the previous ones -, described as "splendid" who are "at the end", or who are the "last". It is not clear if he is talking about exo-planets, dwarf planets of the solar system and/or natural satellites, but he says that they are from 'ha.Avir' (from the air). It would not make sense for him to be insinuating that there are planets or small stars in our atmosphere, but if you remember the theses in 'RS3' and 'Transcendence', you will be able to collect the data on the "regions of the air", which is the way of referring to space closest exterior to our world, that is, what to a certain extent we could define as the first heavens of the Earth. The ancient peoples referred to the space of the stellar Sphere as the regions of the air, and that we could say that it is everything that - including the Earth - encompasses the space that reaches beyond the 12 constellations of the zodiac and the 36 decans.

Whether it refers to bodies in this system, or to the constellations (which I would define as "small" due to their distance), the fact is that it encompasses all these stars. The mention of Moses in Jubilees 2:8-11 does not give too much detail either: "And on the fourth day he created the sun and the moon and the stars, and set them in the expanse of the heavens, to give light to all the earth, and in the day and the night, and divide the light from the darkness. And God appointed the sun to be a good sign on earth for days and for Sabbaths and for months and for celebrations and for years and years of Sabbaths and for anniversaries and for all seasons of the year. And that separates light from darkness [and] from prosperity, that all things may prosper and grow over the earth. These three guys made the fourth day.

וַיַּעַשׂ אֱלֹהִים אֶת־שְׁנֵי הַמְּאֹרֹת הַגְּדֹלִים אֶת־הַ מָּאוֹר הַגָּדֹל לְמֶמְשֶׁלֶת הַיּוֹם וְאֶת־הַמָּאוֹר הַקָּ טֹן לְמֶמְשֶׁלֶת הַלַּיְלָה וְאֵת הַכּוֹכָבִים:

VERSE 16 OF GENESIS 1 says, " ve.iás Elohim et-shnei ha.morot ha.gdolim et ha.maor ha.gadol lememshelet ha-iom ve.et-ha.meor ha.katan lememshelet ha.laila ve.et ha.kokabim », a quote that states that the purpose of the greater Meorah is to lord or govern over the Yom (aeon, day), while the lesser has it over the Lalila (night). This appreciation is not metaphorical, since the creative Logos retreats throughout the universes at all scales, according to the personification of consciousness at the level that has been established. However, the Logos that created the Milky Way and

our Sun, being the consciousness of this Sun Logos who would then create our planetary Logos, who in turn would determine the creation of our particular logos or individualized portions of consciousness. Ra says: "With the fundamental distortion of free will, each galaxy developed its own Logos. That Logos has total free will to determine the paths of intelligent energy that promote the lessons of each of the densities, according to the conditions of the planetary spheres and the solar bodies." (Ra Material, 1981)

How does the Galactic Logos create the Solar Logos and that Solar Logos the Planetary Logos, and successively, creating our being or sub-logos? Awareness. It is always consciousness, since it is a Mind creating within its own dream. Unfortunately, the human of this world and level believes that the things that happen to him are the result of external powers and alien to him. He attributes the good and evil of his particular experience to diabolical forces or the destiny of gods, leaving aside responsibility for his own life. Our Mind ca the reality we perceive; Our thoughts are the precursors of the life that appears before us and in our body. When we are aware that our Mind creates what happens in our quantum field, we stop being victims of the randomness of our irresponsibility, and we take control of our thoughts, emotions and actions. In this way, our Mind begins a new stage, one in which it consciously chooses which thoughts to accept and which to reject; You then deliberately configure the projections that you want to manifest in your life. Your Mind begins with the images in your thinking, and from there come your emotions about what attracts or pushes away, and finally you use your hands and intellect to carry out your desires and dreams.

While man lives in superstition, he deprives himself of his power, he loses the ability to take control over his life; He cedes his life experiences to assumptions and beliefs that he has absorbed from his ancestors or religious influences, and then he justifies himself and his misfortunes by saying that "it is destiny", "they are attacks from the devil" or "they are tests from God". Since that only happens at our Third Density level, as Logos Adam we let the arbitrariness decide what we should be deciding. The consciousness of the Galactic Logos, the Solar Logos and the Planetary Logos, as well as the Dimensional Logos, attract with their thoughts the things that will configure their substance of atomic and energetic levels, as well as their electromagnetic fields, orbital bodies and circumstances. Thus a Logos could be said to create its own body from the substances of Intelligent Energy, as well as the scenarios that will give rise to the manner and methods in which other life forms of other densities will develop in its avatar. In this way, a Logos like that of our planet attracted the consciousnesses that fertilized the earth, alkalized the sea and sowed the seeds of life and the DNA of animal species so that later others would come and produce the conditions for Adamic life.

As far as our solar system is concerned, there are the two key Meorot – the main ones -, as symbols of the times, and then there are the kokabim, which - as we have analyzed - define the rest of the meorot - be it planets or stars -. According to what can be seen in the following quote that says « ve.iten otem Elohim ba.rakia ha.shamaim lehair al-ha.aretz », we observe that they were designated to "illuminate" on Earth, and just as in the text from Henoc, we find the word 'Air', but interpreted as the conjugation of the form 'Aor' (light), in the sense of illuminating, so it would have a double meaning: region of the air and illuminating (let us bear in mind that the stars and the Moon illuminate the night sky - especially Selene due to its proximity and consequently its volume -, although not the Earth, but the Sun does, since given its proximity the photons hit the entire atmosphere, illuminating it, in such a way that they produce the effect of "daylight").

וַיִּתֵּן אֹתָם אֱלֹהִים בִּרְקִיעַ הַשָּׁמָיִם לְהָאִיר עַל־הָאָרֶץ

Moses says, then, in Hebrew characters from verse 17 of Genesis 1: "ve.iten otam elohim ba.rakia ha.shamaim le.ha.air al-ha.aretz", where we can read that he affirms "about the air that is on Earth"; but as I explain in RS3 (Ch. 4 - Governors of Darkness, page 152), Air or Aera is the dimension of the Heimarmene, where the stars that make up the entire extension of the 12 constellations of the zodiac expand. This can make us suppose that the stars of the Heimarmene determine the time-destiny on the material worlds (Aretz). It sounds like the influence of the stars, the so-called horoscope, but it is that "when the river sounds" it is not that an orchestra has drowned, but that the beliefs so deeply rooted and influential in so many peoples of the globe come from somewhere. 2 Enoch 30, for its part, adds: «And guard over the Heavens to illuminate the Day and over the Moon and over the stars to illuminate the Night. And the Sun must go around the whole circle [of the constellations] of the zodiac; and 12 circle the zodiac revolve [around] the Moon; and they act according to the meaning of their names and thunder according to the circle of the zodiac that is in front of them, and [the] law of their hours according to their rotation." (Verses 5-6)

I find it fascinating to see that in remote eras it was known - without advanced measurement equipment - that the sun makes a trip around the region of the constellations, since it is a concept that would be difficult to determine without a computerized system - or a patient and ancient observer. and/or marking points of celestial movements counting the changes century by century -. Yes, because we are talking about very long cycles, from Baktun (400 years) or Dan, from Dan'ha (2,000 years), of zodiacal eras (2,150 years), of complete zodiacal cycles (25,800 years) and other much longer periods, like that of the complete cycle of the Milky Way. Ironically this 25,800 year cycle is the total time for the precession of the equinoxes, and the cycle that many consider corresponds with the Sun's journey around Alcyone - which would be the central star of the Pleiades, where some assume that our Sun is the 8th component of said group -. As in the previous case, Moses reiterates in chapter

1:18 that the role of these king stars is to dominate in Yom and Laila, and to separate between Light and Darkness (you can study the complete thesis on page 161 of RS3, in the chapter on 'The 4 Angles'), seen as something positive:

וְלֹא וֹר וּבֵין הַחֹשֶׁךְ וַיַּרְא אֱלֹהִים כִּי־טוֹב

Enoch's descriptions of the 'Meorot' are clearly a thousand-year-old treatise on astronomy: "Observe all the things that happen in Heaven, how the luminaries of Heaven do not change their path in the positions of their lights and how they are all born and laid (installed), each one ordered according to its season and do not disobey its order." (1st Enoch 2:1) The positions of its lights? According to what this legendary prophet tells us, the Meorot are established in specific places - not in a random order - and their light intensity (magnitude) is deliberately set according to a pattern or prerequisite. In addition to this, Enoch speaks of the Meorot as if they were "consciousnesses", not mere spheres of light (assuming that we are also talking about 'spheres' here), and that they definitely participate in the reality of existence and even of the determinations universals and the right of free will: "Rauel" is "one [of] the malachim (messengers), sacred, the one [who] takes revenge on the world [of] the luminaries" (1st Enoch).

If the Meorot had no conscience or conscious actions, how could they do something that would lead the archangel Rauel to "take revenge" on them, or retaliate? It is clear that the close relationship between the idea of Meorot (luminary, light) and Kokabim (star) with that of Malajim (angel) is so close that it even seems to indicate that these stars are a high manifestation of consciousness and/or a consciousness in around which conscious beings are connected and linked. For those who doubt this relationship, and even the fact that the "heavens" are indeed portions of the universe in relation to planes and dimensions that are intertwined with the physical, psychic and ethereal, Henoc has some very constructive words: «The angel

Michael He took me by the right hand, lifted me up and led me into all the mysteries (secrets) and revealed to me [...] the secrets of the limits of Heaven and all the reservoirs of the stars, of the luminaries, whereby they are born in the presence of of the saints. He transferred my spirit into the Heaven of Heavens and I saw that there was a crystal building and between those crystals, tongues of living fire.» (1st Enoch 71:3-5) Not only does he make a difference between Kokabim and Meorot again, but he says that the angels are in front of these "consciences" that are created, understood in the place where they are created.

The Oahspe is one of the writings that makes this distinction, speaking of "suns" and "photospheres", adding that the original stars are not of this perceptible state of reality, but some are made "physical". Where are stars and planets created? Unless they are produced in another universe and spit out to this one, it seems obvious that there is an interaction of worlds, and this takes place in "the heavens" from one dimensional room to another. In keeping with the Oahspe, photospheres have been produced from the ethereal worlds to illuminate and awaken consciousness in the physical worlds. Oahspe would also clarify that having 3 spheres-states of reality (corporeal, atmospheric or psychic, and ethereal), this type of space bodies would develop between these levels (being the nebulae - the cradle of formation of these stars - of an atmospheric nature, or psychic).

In the 'Book of the Luminaries of Heaven', Enoch describes each sequence, cycle and period, both of the Sun and the Moon, and the establishment they have and the angels that direct them to fulfill their objective: «The Book of Movement of the Celestial Luminaries, the relations between them, according to their class, their dominion and their station, each one according to its name and the place of its departure and according to its months, which Uriel, the holy malaj who was with me and who he is your guide, he showed me and revealed all his laws exactly as they are and how they are observed all the years of the world, until eternity, until the completion of the new creation that will last until eternity. This is the first law of the luminaries, the luminary of the sun, which has its birth at the eastern gates of Heaven and its setting at the western gates of Heaven. I saw 6 gates where the sun rises and 6 gates where the sun sets, and the moon rises and sets through those gates, as well as the leaders of the stars and those who guide them." (1st Enoch 72:1-3)

Later he adds about this "Greater Luminaire": «First there appeared the great luminary whose name is Shemesh (the sun) and whose circumference is like the circumference of Heaven and is completely filled with a fire that illuminates and burns. The Ruach carries the chariot in which he ascends and the sun sets and returns through the north to return to the east and is led so that it enters through that door and shines on the face of Heaven. If the Greater Light has a "circumference" that is equivalent to the "circumference of heaven," this coincides with the revelation of Zephaniah and that of Baruch about the journey of these knights to the heavens, since they define such heavens as "spaces in space" - but in several dimensions -, and they affirm that the "distance" of the Earth from the Sun is like the size of that sky - similar to the Hindu version addressing the same theme -. But what parameter would determine the magnitude or proportions of each sky? Apparently, the planetary sphere or star itself. Oahspe calls the stars of etherea, 'Photospheres', and also agrees that planets, stars or suns and photospheres - or vice versa - are the axis around which dimensions are wrapped, one on top of the other, creating what could imagined as a layer or mantle so thick that seen with spiritual eyes from a great distance, the planetary sphere itself would be almost like a tiny point next to its corresponding sky - or heavens within its sky -.

The role of the Sun, like that of the Moon, would be, effectively, to counteract invisible, energetic and potential forces, preventing chaos and darkness - with all its shadowy powers - from destroying balance and order. For this reason we see descriptions as apparently incoherent as saying that the sun watches over the day and night, when he is the one who determines this. But if we understand that Day and Night are concepts, it makes total sense that the Sun and the Moon have their strategic and decisive role in this balance: « This is the law of the sun's journey and its return, according to which it returns and is born 60 times, so the great luminary that is called the sun, forever and ever. It is the great light that arises, named after its own appearance, as the Lord has ordained. Just as it is born it hides, without decreasing or resting, but traveling day and night; and its light shines 7 times brighter than that of the moon, although when observing them both they have the same magnitude. (1st Enoch 72:35-37) The Sun projects photons, which are packets of information - consciousness -, either directly or indirectly (as it does with the Moon).

Finally, and not to go into much more detail with this particular point, I will comment that the Moon is a reference "satellite" with respect to the constellations of the zodiac and the Earth, an electromagnetic influence for the energy fields of the Earth, and all this, in relation to the control of the powers and forces that govern chaos (since the magnetic, photonic and gravitational force - being related - are blockages for "doors" or portals that open and close in their passages, convergences, alignments, shadows (eclipses) and movement, like a great gear clock, to allow the transit of one or another powers of the invisible planes of reality that maintain the balance of the cosmos): « After this law, I saw another law, which deals with the little luminary, whose name is moon. Its circumference is like the circumference of Heaven and the chariot in which it rides and the light is given to it in measure; and every month its birth and

154

its setting are modified; His days are like the days of the sun and when his light is full, it is one seventh of the light of the sun. (Chap. 73:1-3)

If the Moon also has a circumference similar to that of the sky, it means that it belongs to the same sky that is being talked about or, more probably, it has a field of extension that in proportion follows the same sequence of the spiritual worlds that are in it. the higher planes of the "physical" spheres. We can take an example of all this from the Pistis-Sofia treatise, where Valentino allows us to know the words of the master Yeshua about the celestial battles: «And their angels, and their aeons, and their archangels, and their archons, and their gods, and their lords, and their forces, and their luminaries, and their ancestors, and their triple powers, saw that I was infinite light, to which no kind of light is alien.» (Chap. 3:20) In this manuscript from the Nag Hammadi Library it is read that Jesus describes all the spaces of consciousness and the envelope of spatial bodies as 'spheres', and also speaks of the "firmament" as a spherical space that is limited at the height of what could be interpreted as the limit of this solar system; then, another sphere would be the one that encompasses the previous one, being something like a space where other stellar systems reside, and on this, another sphere, which would be the circle

of the zodiac constellations.

In addition to talking about Mars, Mercury, Venus, Jupiter and Saturn as consciousnesses that are interrelated with the 12

constellations and the 36 decans, Yeshua in this text is very emphatic when stating that all these celestial bodies are beacons or starting points of the spheres of existence and destiny, everything being operated from different dimensions: "And it would have been a long time before the archons of the aeons, and the archons of Destiny, and of the sphere, and all its regions, and its heavens, and their aeons would have been destroyed. (Ev. Valentino 5:7) The 2nd book of Enoch speaks of "the little ones" that are apart from the 7 great bodies. However, we have two luminaries and 5 cold stars – as they would call them in ancient Hebrew – and a lot of lower bodies. In the solar system alone there are officially 168 moons, apart from another 6 that revolve around the other 5 bodies that have been called dwarf planets of our system (Ceres, Pluto, Haumea, Makemake and Eris). These bodies seem like a guarding entourage around all the planets in our neighborhood, except for those closest to the Sun: Mercury and Venus. This entire structure of bodies was established under a larger structure, and each of these celestial bodies has a whole hierarchy at various levels.

The Intermediate State

These planetary bodies and their companions support an interwoven network of dimensional channels that control the embodiments of chaos on these planes of reality. In the same way, this web of invisible fields between the planets and their moons - which start from the toroidal foci of consciousness - contain the projection levels of the levels of consciousness, so it could be said that the planets and their Moons, as well as the stars, are the beacon, or reference base from which the various planes of reality through which consciousness travels at the lower levels of perception are intertwined. Therefore, the states that we would call meditation, sleep, death, astral, etc., are carried out through these channels, and in them, according to the measure of the level of evolution of consciousness and the particular learning cycle of each consciousness. (soul). This is what Masters Arten and Pursah call "states between lives," or which Master Seth encompasses in the context of states outside of wakefulness.

This means that the events conceived by the Mind as "death" are projected into consciousness when it leaves the body in the intertwined planes of these worlds and their dimensions. In this way, the projections of the unconscious on the archetypes that hell would have are created in forms, figures and scenarios, as well as demons, which take on the characteristics that the Unconscious Mind considers they should have - according to the archetypal perception of guilt and belief that has absorbed -. In this way, the Intermediate Regions – as the apostles called them – project the scenarios that the Collective Mind has created due to its beliefs and social traditions, its myths, fears and primitive configurations. In this way, the places would be created that the archons - or diffusions of the erring mind of consciousness - would project to punish the soul that believes it deserves punishment. There the souls experience what they believe is their purification - as long as they have not learned forgiveness - and therefore they must return to a body until true forgiveness cuts the Samsara (cycle of incarnations).

It is in these regions where the Mind conceives its own fears and guilt, and it exists because it is collectively manufactured by the society that unconsciously believes in it, as it craves punishment for its feeling of guilt. In this way the idea of a Jalukam - or Jaluham - is understood - giving the water of oblivion to the souls, by order of Adamas the Tyrant (Tzabaot Adamas) - whose servant he is - so that the soul returns to incarnate without remembering, and In this way, your conscience will learn the lessons of forgiveness that will make you wake up from sleep. Yes, this is also why there is the idea of Abiuth and Carmon (servants of the demon Ariel), who take the soul to see the Intermission for 3 days before making it suffer what the soul believes it should suffer, in periods that stop the terrestrial stages. It could be months to decades.

According to the state of guilt, this apparently individualized consciousness travels through the 4 states of greatest suffering, which are nothing other than its worst perceptions of lower vibration, from hell to the Outer Darkness, passing through the tortures of the places of the middle path. or the great chaos with all its demons. Who created this? The man himself. Their fears take shapes and forms, they produce scenarios, just as happens when we dream, and the forces of consciousness around the being adopt the external form that the person wants or needs to see. From there were born the comings of demons from hell, Proserpina (Persephone), Caldauoth, Cernunos, Ereshkegal, the god Hades, the demon Surt and many other fantasies of the traditions of the people. But in another passage later I will go into greater detail about this.

Day Five

If in itself the appearance of plant life is fascinating, how much more is animal life? Accepting that plant eukaryotic cells could not appear alone, it is conceivable that the same would replicate with animals. The prophet Enoch tells us about this in just one verse, saying: "And on the Fifth Day I drew out of the sea and there came forth fish and birds of various kinds and every reptile that crawls on the Earth and what goes on four on the Earth and flies in spirit male and female in the midst of them, and every soul that breathes for all life." (2nd Enoch 30:8) According to ancient texts – and modern science seems to agree with them – the first forms of life in our world were androgynous creatures, that is, hermaphrodites, each possessing both sexes: male and female., at once. This 5th era, which seems to be the general period of the Mesozoic, saw aquatic, aerial and "reptilian" life. It would seem that this supports the hypothesis of evolution, but what this could adduce is that the birds appeared before the reptiles, or that they would have arisen "from the water", which does not have to be correct.

All things came from a hermaphroditic origin, and later their species were separated into males and females, starting with plants and then marine creatures - not mammals -. This is a paragon of what is below with the root emanations of the higher realms, and therefore, the example of what they must return to: "J said: 'When you make the two into one, you will become the son of Adam, and when you say, 'Mountain, get out of here!', it will move.'" (Ev. Tomás 106/104) In reality, all species intrinsically have our dual part (males the female part, and females the male part). Enoch wrote that he was told by the Creator that life was taken from the sea in this period, and according to what was carried out, these life forms were successfully produced in large quantities in the ocean – and /or in fresh water -, this aquatic world is a clear example of life forms that exchange their sexes and even their reproductive roles.

What Enoch's text also seems to imply is that the same biological production project succeeded in creating "birds". The term 'Aof', although it is translated as 'bird', gives rise to the word 'Meofef' (fly), in the sense of the action or exercise that birds perform: flight. For this reason the Hebrew word 'Kanaf' (wing) derives from the same cognate. Genesis tells us that reptiles appeared on the Sixth Yom, so birds could not have come from reptiles – as the Natural Selection hypothesis first proposed by Charles Darwin maintains. Likewise, the texts maintain that the first forms of life came out of the "sea", so it is assumed that what they are trying to imply is that the "experiments" of life were carried out out to sea, and resulted in the forms of life. marine and, later, those that would fly the skies (first androgynous and then separated by sexes that complement each other for reproduction and maintenance of life).

In the case of Moses - both in Genesis and in Jubilees - it is more explicit in this story: «And on the fifth day he created great monsters of the sea in the depths of the waters, for these were the first things of the flesh that were They have created by their hands the fish and everything that moves in the waters, and everything that flies, the birds and all their kinds. And the sun rose above them to prosper (them) and, above all that was on the earth, all that sprung from the earth, and all the fruit trees, and all flesh. These three types He created on the fifth day.» (Jubilees 2:11-13) Here he tells us about something that Enoch deals with in another section, and it is about those such "sea monsters" that the Septuagint translates from Genesis 1:21 as 'Kiti'. From the Greek word 'Kiti' comes 'Kitei', and from there 'Kiteos', which derived from Latin 'Cete' and 'Cetus', and into Spanish as 'Cetáceo'. For this reason there are references to the Kitim (Kiteans) in the Tanakh to evoke the peoples of the north of the Mediterranean Sea. This designation was later used by the Jews to refer to the Roman people.

It is common to hear in ufology stories that beings from other worlds brought life to Earth, the first creatures with consciousness - which according to the defenders of these thesis would have been brought from Sirius - the cetaceans: dolphins and whales. Another aspect is that which encompasses the character of giant reptiles in these creatures. The word that Moses uses in Genesis is 'Taninim' (whose singular is 'Tanin'), which is usually translated in the Bible as 'dragons', although 'Tanin' in modern Hebrew means 'crocodile', and the form 'Tan' – plural 'Tanim' – is 'jackal'. What is clear is that these statements coincide in saying that these forms of life were the first "meat" forms that were produced; It also includes the marine biodiversity of so many aquatic organisms that exist, and that were also created, as well as the rest of the things that live in the ocean, in underground aquifer galleries, in lakes and in rivers. This passage from Jubilees is also important, because it clarifies that such "fliers" were not simply birds, but "zbubim" (flies, flying insects). We had also said that in the Third Yom plant life was produced, but that the

Sun did not enter the scene until the following cycle (the Fourth), and it is here where it says that then! The king star crossed the sky and made everything "prosper."

So that plants and trees, despite existing long ago, only really began to bear fruit from this moment - so before they must have been in some kind of "vegetative" state or dormancy, without yet producing fruit -. Consequently, it is evident that it was not necessarily seeds that were scattered on the ground, but vegetation already prepared - previously arranged - for when the passage of the Sun arrived. This suggests whether the plant life of that time was far from resembling the current and have a longer-lived and quasi-immortal appearance, or was even grafted from laboratories and hydroponic crops or artificial greenhouses.

וַיֹּאמֶר אֱלֹהִים יִשְׁרְצוּ הַמַּיִם שֶׁרֶץ נֶפֶשׁ חַיָּה וְעוֹף יְעוֹפֵף עַל־הָאָרֶץ עַל־פְּנֵי רְקִיעַ הַשָּׁמָיִם׃

IN GENESIS 1:20 WE read: "ve.iamer Elohim ishratzu ha.maim sheretz nefesh jaiah ve.of iofef al-ha.aretz al-pnei rakia ha.shamaim", which translated means that the sea was said to "flow" a swarm of "living souls" and flying birds on the earth, in front of the celestial firmament. It is important to constantly repeat that all these words and phrases have a double representative component, since in addition to their direct meaning they have a symbolic nuance, wanting to show an analogy with the work of angels and the human personality. Thus, all things in the world of phenomena are the image of heavenly things and of the psyche. Now, these "living souls" had not yet been described, and they become what in the Hebrew language is 'Nefesh Jaiah', that is, a form of life that depends on breathing (of air), that has a soul, the privilege of existence and movement in a sphere of reality (in his case, the physical world).

Just like Enoch, Ezra speaks of these "monsters" and the life forms that were created in that era: "On the fifth day you said that in the seventh part, where the waters were gathered, that there should [be] living creatures, birds and fish, and so it happened. For the dumb and lifeless water gave birth to living beings at the commandment of God, [so] that all people could praise your wonders. Then, you ordained two living beings, one you called [Be]he[m]o[t], and the other Leviathan; And you have separated one from the other: from the seventh part, that is, where the water was gathered together, could not maintain both. And to [Be]he[m]o[t] you gave a part, which was withered on the third day, that they should live together in the same part, in which the thousand hills: However, to Leviathan you gave the seventh part, namely, the humidity, and you have saved it to be a devourer of whomever you want, and when.» (2 Ezra 6:47-52) Unless Behemoth and Leviathan were reptilian mega whales, these creatures do not seem normal. Doing an in-depth study on this matter, one could conjecture that these two "beasts" constitute a group of "creatures" under one spirit.

For this reason, speaking of the end times, Henoc was told that "on that day two monsters will be made to come out separately, one female and one male. The female monster is called Leviathan and lives at the bottom of the [great] sea above the source[s] of the waters. The male monster is called Behemoth, it rests on her chest in the immense desert called Dondain, to the east of the garden inhabited by the elect and the just, where my old father was taken, the seventh since Adam, the first man created by the Lord of the spirits." (1st Enoch 60:7-8) If these two monsters are reserved for the end of time, why does the Apocalypse of John not speak of them? According to Enoch, these beasts will devour meat, the writer implying that this "meat" is human, but on a metaphysical level they identify egoic thoughts. Apocalypse the only thing that speaks in analogy with something like this is regarding two powers that will subdue the world for a few years. If we look at the similarities - as can be seen in the 'Sakla Rebellion' books – Behemoth (meaning "the beast") is a male creature who - according to Jewish tradition - rules all land animals; For its part, Leviathan, is a female creature that, according to the same tradition, directs aquatic life forms.

The most coherent correspondence at the level of form is that Behemoth could be related to the Beast of Revelation, and Leviathan could be related to the Dragon of the same story. The question would be, how come these creatures were created so long ago, especially if their function is for the "final" era? The Vedic culture of India tells in its sacred writings that at the beginning of time - when 'Lilah' (or 'Lika') began - the Creator of the cosmos (Brahma) faced the sovereign powers of chaos, such as Vala and Vritra, the first demon-gods, whom he defeated, but also - manifested through Vishnu - he fought against the titanic Rajú (the 'Attacker'), and divided this creature into two: Rajú and Ketu (the one who was "disconnected" and wandering through space). The attacker is called Rahab ('the Proud One') in ancient Hebrew history, and it is told of him in the psalms: "You broke Rahab like one wounded to death; With your mighty arm you scattered your enemies. (Ps. 89:10 - KJV 95) This victory is also mentioned in the books of the prophets: «Awake, awake, clothe yourself with power, arm of Yaheveh! Wake up like in ancient times, in centuries past! "Are you not the one who tore Rahab to pieces, the one who smote the dragon?" (Isa. 51:9)

If we talk here about celestial battles, we have to add to our analysis the fact that the Enuma-Elish relates that the "evil spirits" who were on "the Earth" - or solar system - in the primeval era, were forced to divide, becoming one of two parts that survived an astronomical impact (which evoked a war between supernatural beings). Those who were thrown into the planetary sphere and those who were thrown into space came to be considered the two spirits of evil that had to be kept separate, since when reunited again, they would procreate chaos. So, following the narrative, we find that on the Fifth Yom these creatures were created, and since it does not tell us when they were separated, we have to assume that it was in that same cycle. Consequently, one could conjecture that at that time some type of war would have occurred already in physical-psychic canons, not between humans (which did not yet exist in Towsang), but between "intelligent" beings. Several of the Nag Hammadi manuscripts talk about celestial wars and their seasons, saying, for example, that some time after chaos and its powers had appeared, there were several battles in these heavens for their sovereignty, and the evil powers were divided and subdued.

The great similarity between these stories makes one imagine a somewhat fantastic - but ordered - story of how various conflicts were occurring in the heavens, and the Earth also experienced them, even at the time when organic life and oviparous creatures were created in our world (and possibly the war issue was related to the plans for the creation of life and conditioning of this planet). But what do Leviathan and Behemoth mean on the level of Mind? Why shouldn't they join? Leviathan and Behemoth are the dual aspects of Mind, the misperceptions of the Left Hemisphere and the Right Hemisphere. They symbolize the distortions of each aspect of duality, or put another way: the polarized parts of Yin and Yang. In reality, the masculine and feminine aspects are spiritual concepts of the virtue of oneness, but in the ego they develop as polarized opposites, inclined to the most dual of the idea of separation. A very simple example of them is to see machismo or Behemoth thinking, and ultra-feminism, or Leviathan thinking. Would it change anything if they joined? If their idea is of separation they only legitimize the belief in duality by implying that evil is necessary for good to be right, or that death and life are inseparable. In reality, both concepts are illusory, since they are part of the dream of the individualization of the Mind from the source of Everything.

וַיִּבְרָא אֱלֹהִים אֶת־הַתַּנִּינִם הַגְּדֹלִים וְאֵת כָּ ל־נֶפֶשׁ הַחַיָּה ׀ הָרֹמֶשֶׂת אֲשֶׁר שָׁרְצוּ הַמַּיִם לְמִינֵהֶם וְאֵת כָּל־עוֹף כָּנָף לְמִינֵהוּ וַיַּרְא אֱלֹהִים כִּי־טוֹב׃

MOSES RECOUNTS IN VERSE 21 of Genesis 1: « ve.ibrá Elohim et-ha.taninim ha.gdolim ve.et kal-nefesh ha.jaiah; ha remeshet asher shertzó ha.maim laminehem ve.et kal-aof kanaf laminehu ve.ira Elohim ki-tob », speaking of "creating" great dragons and every living soul. Sometimes he says to "create", others that he commands to "produce" and others that a certain thing "comes out", each case being different from one another, since on this occasion it is clear that the gods (elohim) create, while in the other situations they "command" - or who is over them - that "something" or "someone" carry out functions of "producing" based on the elements already created - or simply that the elements themselves, or existing mass, is transmuted to establish new scenarios -. Now, why would he create "big" dragons? It is notorious that he refers to the fact that he created two sides to lessen the power of these forces, but he also speaks on the earthly side of the creation of great dragons. Celtic records tell us, saying that the "Earth was clothed in the mantle of the lady of green, grass covered the face of the earth. The waters produced fish and the creatures that move about and writhe and coil in the waters, the serpents and terrible-looking beasts that were of old, and the creeping and crawling reptiles. There were things stomping around, and dragons in hideous form cloaked in terror, whose great bones can still be seen. Then came out of the belly of the earth all the beasts of the field and the forest." (The Book of Creation. The Kolbrin).

Unequivocally these are direct and indirect allusions to dinosaurs. In Genesis a difference is made – or so it seems – between two groups: taninim ha.gdolim (great dragons) and nefesh ha.jaiah (living creatures with souls). Moreover, these living souls are defined as "all", that is, on the one hand there were these mega reptiles, and on the other, "all" living souls. Does he mean that the other big monsters weren't creatures with souls? In fact, these living souls are categorized as those first oviparous, namely, forms that crawl on the subsoil or sea bed, and the "winged" birds. But are there birds that are not winged? If a characteristic of birds is understood to be their wings, why do you highlight it, and why not before? It is probably because it speaks of an order of creation, where fish and INSECTS (the simplest complex life forms) were produced first, and just after "laboratory" work developed dinosaurs, molluscs and crustaceans, and the birds. It might be thought that the great jungle expanse of mainland was so depopulated of animals that testing the design of the first reptiles there would be ideal, although due to its size it got out of hand (the other forms of life that would accompany them - the birds - they would be flying, out of reach).

וַיְבָרֶךְ אֹתָם אֱלֹהִים לֵאמֹר פְּרוּ וּרְבוּ וּמִלְאוּ אֶת־הַמַּיִם בַּיַּמִּים וְהָעוֹף יִרֶב בָּאָרֶץ:

Genesis 1:22 says: "ve.ibraj otam elohim leemor pru ve.rabú ve.malú et-ha.maim ba.iamim ve.ha.aof ireb ba.aretz", referring to the fact that these forms of life had to fructify and multiply in the sea, while the birds did it on Earth. But why only birds? Aren't dinosaurs also supposed to be multiplying on the dry surface? This makes one wonder if the great dragons to which he initially refers were indeed just the oceanic monsters, such as megalodons, plesiosaurs and other species that were the giant companions of whales and other life forms that are still studied in cryptozoology, such as the giant squids and others that are spoken of in Hindu and Japanese myths - to cite these two examples only -. The stories and engravings on these specimens are seen in many places, and it is known that their size was colossal, mixing with the stories about the fights between the gods, such as when Horus fought with Set (in Egyptian stories), which is the myth Greek of the confrontation between Zeus and Typhon. In his book 'Flying Serpents & Dragons' (1990), RA Boulay theorizes on the very direct relationship between the appearance of dinosaurs and the wars between reptilian-like demonic beings, using all sorts of sources, including the Bible and the Hagadah itself. (a text from Jewish tradition).

The use of analogies is key to understanding both levels of information received, as I have already said, since the messages are given with similarities between the celestial and the earthly, showing the etheric and the phenomenal as one in the image of the other. Looking at the accounts of these "monsters" written by different characters in Hebrew history, we find that they say that these sides represented the ancient dragon, some as 'Leviathan the Fugitive Serpent' and others as 'Leviathan the Tortuous Serpent' (Isa. 27). Yeshua tells about the celestial combats precisely describing them as "battles against dragons" - which is a classic that appears in practically all prehistoric stories or fables -: «And when the power of the triple power had descended into chaos, he found Pistis Sofia. And the force with the face of a lion, and the force with the face of a serpent, and the force with the face of a basilisk, and the force with the face of a dragon, and all the forces of the triple power surrounded Pistis Sofia, wanting to snatch away for the second time his forces. And when they tormented and afflicted her, she turned again to the light. (Ev. Valentinus 20:10-12)

The account of the Enuma-Elish (Sumerian epic of creation) makes us aware of all this in many details, but clearly through figurative descriptions: «They themselves [were] gathered together and at the advancing side of Tiamat; They were furious; they devised mischief without rest Day and Night. They prepared for battle, fuming and fury; They joined forces and made war, Umm-HUBUR [Tiamat] Maker of all, Made into invincible weapons of addition; she engendered serpent-monsters, sharp-toothed, and merciless [...] With poison instead of blood, she filled their bodies with ferocious snake-monsters, dressed in terror. With the splendor that clothed them, he made them tall in stature..." This Sumerian description encompasses all of the above, speaking of what appear to be demonic creatures of different species, all of them engaged in a fight in the solar system (since Lahmu, Kingu, and other deities described throughout the conflict, are the names that some experts assume that the planets and moons of our solar system received in Mesopotamia).

The aforementioned tablet continues to say: "Whoever looked at them, terror took hold, their bodies reared up and no one could resist their attack. He established vipers and dragons, and the monster Lahamu, and hurricanes, and mad dogs and scorpion-men, and strong storms, and fish-men and rams. They carried cruel weapons, without fear of fighting; His orders were powerful, he could not resist them; before this appearance, of great stature, he made eleven [types of] monsters. Among the gods who were her children, to the extent that she had given her support, she exalted [Kingu]; in their midst he raised him to power. To march before the forces, to lead by host, to give the signal for battle, to advance to aggression, to direct the battle, to control the fight, To him he entrusted; in expensive dresses that she made him sit..." On the level of form, these show to be beings of various dimensions coming into conflict with the forces of light; creatures that folklore has variously called demons.

Part V

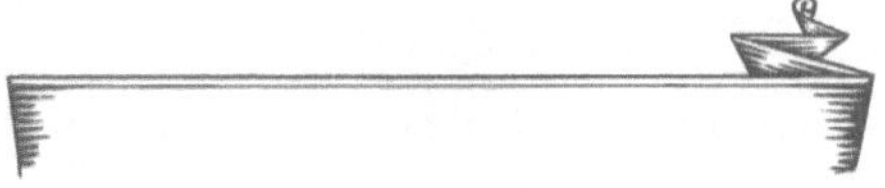

THE AVATAR WAS CREATED

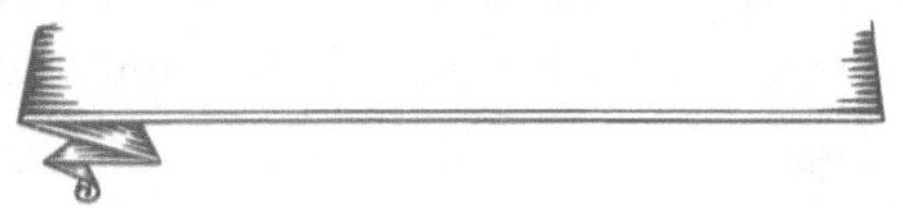

«And I am the one who comes from what is total.
They have given me some of my father's things.
Therefore, I say that if one is total, he will be full of light,
but if one is divided, it will be full of darkness »
(Ev. Tomás, 61)

Days Six
The last eon of the creation process described in the Genesis of Moses is the 6th, the stage in which the terrestrial reptiles and mammals appear, and also the human being, the last to manifest and for whom it seems that everything was arranged and conditioned.. About this period Ezra wrote: «On the sixth day you gave a commandment to the earth, to give the beasts, the cattle and the reptiles, before and after these, also Adam (man), whom you made lord of all your creatures. Everyone comes from him, and also the people you have chosen. "I have spoken all this before you, O Lord, because you made the world for our sake." (2nd Ezra 6:53-55) According to these words, the order was given to create mammals of two genders and reptiles - previously, and deliberately, to the arrival of man -, and concluding with another reaffirmation that everything was established so that humans would be lord over everything and have control over everything.

Jubilees contains: "And on the sixth day He created all the animals of the earth, and all the livestock, and everything that moves on the earth. And after all this he created man, a man and a woman he created them, and gave him dominion over everything that is on the earth, and in the seas, and over everything that flies, and more beasts and more of cattle, and above all that moves over the earth, and over all the earth, and more than all this he has given dominion. And these four guys he created on the sixth day. And there were a total of twenty-two types. He and all his work [was] finished on the sixth day; what is in the heavens and in the earth, and in the sea and in the depths, and in the light and in the darkness, and, in everything. (Ch. 2:14-16)

What Animals Are You Referring To?

Here he tells us about 'Jayat' of the Earth, 'Behemat' and the rest of what "moves", possibly meaning "what crawls" (in Hebrew 'ha.Remeshet'). The word 'Jaiat' (living) is the conjugation of 'Jaiah' (living), whose root is 'Jiah' (life), which is why it is understood as a living "being". Behemat is a synonym, more explicit according to the animal or animalistic character, but which is used to define beasts and, to a greater extent, for livestock (hence why some interpret it as a domestic or pack animal). It is very strange that he always makes a distinction between 'Jaiat' or 'Behemat' to encompass terrestrial animals (with the exception of reptiles, and, by extension, it is understood that amphibians as well). This curiosity suggests that 'Jaiah' would be the equivalent of the Latin 'animam' (animal, animated being), while 'Behemah' would be the equivalent of the Latin 'beast' (wild animal, irrational creature), which includes wild animals that They have become domesticated. For our society, a generality as simple as saying "animals" would have been understood more than enough, so why emphasize this as if it were two different groups? Possibly because it is characteristic of every ancestral culture to know how to differentiate between coexistence animals and dangerous animals, as well as carnivorous creatures and herbivorous (or omnivorous) creatures, without counting on the euphemisms to refer to intelligent beings, due to their attitude or virtues /defects.

As we can find in many manuscript information and hidden meanings in the scriptures, there are constant allusions to various humanities that have existed, both on this planet and in other star systems. Those in whom the spark of consciousness was would be those who preceded humanity in our world, and some of whom arrived here thousands of years ago. Those who arrived were not the only ones, as there was another "regressive" group – or SAS – who only wanted their supremacy. The Jaiat and the Behemat will be the representative way of identifying these two groups. However, by mentioning conventional humanity before, it suggests that the episodes that occurred in our world with the evolution of life is, in effect, an analogy of the processes of development of consciousness in this universe. Consciousness began to be experienced in various ways in non-simultaneous space-time continuities, but with two parameters in common: Consciousnesses already chosen as gods, and consciousnesses that would begin their path of ascent from the lowest seats until reaching the top.

This implies that the vast majority of the gods and angels that have appeared in our world have already gone through the levels of development through which we are rising. Regardless of the level of vibration-dimension in which they now find themselves, at the time they went through almost all the phases that we live through, except for nuances that are not necessarily identical due to the fluctuations that exist in the variants of the space-time continuity, due to the Mind's own decisions. This is so because each determination of the Mind opens a new line of space-time continuity of different probabilities-possibilities, both in the manifestation of sleep, or material and psychic planes, as well as in the variants of non-waking mental states. While it is true, all the probabilities-possibilities of the infinite options of the space-time continuity of each decision configure "new" scenarios, in reality everything is predestined. Father One already knows the entire history of all the possible scenarios and how they will play out within the screening of this "movie." The Holy Spirit also knows the script from beginning to end, and therefore plays with all the possibility-possibility variables and the various perceptions of the space-time continuum to direct the Mind back home.

וַיֹּאמֶר אֱלֹהִים תּוֹצֵא הָאָרֶץ נֶפֶשׁ חַיָּה לְמִינָהּ בְּהֵמָה וָרֶמֶשׂ וְחַיְתוֹ־אֶרֶץ לְמִינָהּ וַיְהִי־כֵן

Moses says in Genesis 1:24, "ve.iamer elohim totze ha.aretz nefesh jaiah laminah behemah ve.remesh ve.chaito-aretz laminoh ve.iehi-ken." As in the other passages, he tells us about 'nefesh jaiah' occurring on Earth, which he defines as 'behemah' according to all its species, 'remesh', from all its species, and 'chaito-aretz', according to their species. Some have conjectured that this reference makes a distinction between the levels of consciousness and development of these creatures, and/or greater differences, such as the very existence of other "beings" of an intelligent nature, on a par with reptiles and beasts. Seen in an analogical way with the spiritual, it could be assumed that, just like the animal forms, there existed, either in this dimension or in others in our sphere, 3 types of beings with consciousness higher than that of animals. This conjecture is reinforced by the study of ancient manuscripts that speak of "mysterious" beings that cohabit with wild life forms, but live – or lived – in areas far from the populated centers of civilization. In his case, NH II, 5 (Treaty on the Origin of the World, from Nag Hammadi), says that "all plants germinate on the earth according to their species, with the seed of authorities and their angels. Then, from the waters, the authorities created animals, if any, and from reptiles and birds according to their kind, with seeds from the authorities and their angels. (Verses 21-28)

The reference to "seed" encompasses both the vegetable grain and the "seed" or "semen" in a sense of animal reproductive biology, which may suggest that these authorities and their angels used their own genetics for the elaboration of life in this planet. Trying to create life in laboratories and transfer it to conditioned environments taking our molecules, tissue and cells as experimental samples might sound strange a century ago, but thanks to the study of the genome and the verification of cloning and hybridization, and even more advanced experiments, We can theorize that we ourselves could already be able to live on other planets by using our genetic material and adapting it to sustainable environments. A certain study time, necessary resources and tests in the field could make ourselves "creator gods" of life in other worlds.

If we look at the deep and wise Egyptian mythology we can see that the idea of animals and gods were linked. Each of the 44 nethers (nome-gods) were linked to an animal: they were represented with a zoomorphic figure. The first to be referred to were frogs and snakes, as well as crocodiles; with Amon there was the ram, with Horus the falcon, with Thoth the ibis, with Seth what seemed to be the giraffe, and they followed: hippos, beetles, reptiles-lizards, fish, gazelles, cows and bulls, lions, jackals, dogs, cats, baboons, owls, "ostriches", wolves, turtles, horses, mongooses, incneumons, storks, cranes, cobras, herons, snow goblins, monkeys, hares, geese, scorpions, elephants, and many others. These animals have been linked to the stars (see RS3, page 76), to the extent that there are close relationships between stars and animals, as well as other elements, minerals and plants, even representing constellation configurations. For example, we see the Passover symbol in association with the zodiac on several occasions, one of them in the 12 elements that were ordered to be consumed: 2 loaves, 7 one-year-old lambs, 1 calf, and 2 rams. The 7 lambs of a year are the same as 7 cycles, since the year represents 365 minor cycles. The Greek, Egyptian and Sumerian zodiac has 8 animals and 4 figures, but even the Chinese has the 12 annual animals (the Mayan has 13, since the addition would correspond to Ophiuchus).

On the other hand, we must understand that consciousness has taken the processes of time to become conscious of itself randomly in the cosmos, focusing on each node of conscious energy, where, while creating, it experiences itself and self-consciousness. known. In this way, even beings who today – or in past eras – were gods, would have come from experiencing the perception of dreams through the most essential elements of the various scales of evolution of consciousness. Thus - as Ra explains in the 'Material of Ra' - the first state of experience of this octave would be that of the First Dimension, which is from the inert states to those of energy in conscious activation in the elements. This state would move from simple elements - such as minerals - to the consciousness of Intelligent Energy that would activate Air and Fire, who would provide the same potential energy of consciousness to the elements Water and Earth. This state of focus of consciousness of the Logos on matter would activate it to enhance the mobilization of atoms and molecules towards a state of active and creative consciousness. This is clearly visible in Moses' Creation story, where light (fire) and spirit (air) intervene on matter (water and earth) to give it shape and configuration.

Subsequently we find ourselves in the Semuan (or Shem-Mu-An), or primal state, where consciousness is already a planetary sphere personifying a Logos, which becomes conscious to produce the Second Density level of consciousness, where the seeds now They acquire mobility and action, that is, they become activated as "nefesh jaiah" (living soul), appearing cells (in fact, a sperm is the animal "seed" of reproduction-conception). It could be said that corporeal worlds are created by a kind of "vapor" from the vortex of consciousness that condenses, and its friction generates heat and melts, becoming a globe of fire in the sky. It is then a newborn world that is put into orbit. Its next phase is the Se'Mu, where the Semuan is the combination of the less dense elements around it, which go on to make up the oxygen on the surface of the Earth, thus giving rise to the gestation stage for the animal kingdom. in the middle of the darkness.

Consequently, the next stage arrives, which is the Ho'Tu phase – or sterility -, which is followed by the A'du phase, or death. After this stage, the phase suitable for us finally takes place, which is the so-called Uz phase, which fulfills the function of taking us spiritually to the invisible realms. Ergo, the planetary consciousness of this terrestrial Logos pushes its sphere of energy and matter, raising it towards the second state of density, producing algae and grass, continuing its activation towards creating plants, shrubs and trees, while other cells are modified to be designed forms. of complex life, that is, multicellular. The attraction of the mind of the Planetary Logos would offer the sphere the opportunity to be cultivated and germinated with the help of scientists from other systems that would have already concluded this state, or Semuan phase. They would have started by putting spores on comets and asteroids that would later carry our world along the long paths of orbital transport caused by gravitational influence.

In this way, the first cellular replicating forms would arrive deliberately from meteorites scattered from other planetary systems to alkalinize our ocean and acclimatize our atmosphere. When the opportune moment arrived, a group of astronauts, experts in bioengineering, would come to study the conditions that their instruments would have discovered to be propitious for organic life, and they would have spent some time carrying out experiments on the biological conditioning of some cell species. This replicating capacity with which the cells and spores were endowed would have allowed them to readapt to various environments, and would have been the precursor state of a fascinating garden of life, given that the energetic capacity of life of this Planetary Logos among various worlds similar to the our in this interstellar region. In this way, our sphere would become a garden and an area suitable for the development of a wide range of life forms, not only biological, but also psychic.

As the planetary sphere was configured as such - in the Semuan period - the toroidal energy also adjusted the Atmos levels to configure optimal scenarios for beings of a higher state of consciousness. In this way, it forms what is the First Heaven of Earth, which does not merely include the gas atmosphere up to the stratosphere, but goes far beyond the magnetosphere and the lunar orbit. That First Heaven are states of vibration of other levels where "plains" - or Plateau - are established, in which the life-experience of multiple consciousnesses develops that are not experiencing their development at that moment under an incarnated state, that is, that They are not on a Corporal plane. Said First Terrestrial Heaven was defined in Egyptian hieroglyphics under the sound 'D'at' or 'Duat', and it was said that it was watched over by the neter 'Hor', or 'Her-Ur', better known in the West as Horus. The First Heaven of Earth is associated with the First Loka - or first world - in the Vedanta worldview, and given its circumference, it is understandable that its size was said to be equivalent to the distance from Earth to the Sun.

וַיַּעַשׂ אֱלֹהִים אֶת־חַיַּת הָאָרֶץ לְמִינָהּ וְאֶת־הַבְּהֵמָה לְמִינָהּ וְאֵת כָּל־רֶמֶשׂ הָאֲדָמָה לְמִינֵהוּ וַיַּרְא אֱלֹהִים כִּי־טוֹב׃

Genesis 1:25 tells us, "ve.iás elohim et-chaiát ha.áretz lamináh ve.et-behemáh laminah ve.et kal.remésh ha.adamáh laminéhu ve-arán elohim ki-tob", highlighting that the determination was made correctly and had a positive result. However, here he begins to use a word that he had not used until now: 'Adamah'. This word does not make sense to apply if man did not yet exist, since this designation defines the name of our world because of humanity. Adamah is the feminine form of 'Adam' (man), and its root is part of the structure that formed the Coptic and Koine word 'Adamas' (diamond). In my opinion right now, this can only suggest one of two things: either there were already humans in this world since then, or our world was already "baptized" or "founded" to welcome the human race. This could be accompanied by a third option: that he is speaking at a psychic level, not necessarily at the level of form. Strangely, Enoch is the only one of these sources that does not agree that land animals were created or produced at the beginning of Yom Six, but rather includes them all in Yom Five, possibly towards the end of this and the beginning of Yom Six.

This reinforces my thesis that we are not simply talking about animals, but about another type of intelligent consciousness. In addition, it also supports my idea that the texts make a clear differentiation between the levels of evolution of planetary consciousness. Therefore, the second level of density – or Second Dimension of vibration – would have the entire range of plant and animal life forms, which are only aware of their environment in said cycle. However, between the Fifth Yom and the beginning of the Sixth Yom, the levels of consciousness of the entire range of the electromagnetic field of orange energy (Second Density) would have been reached. For this reason, the supposed animals of the Sixth Yom could actually be the creatures that are classically interpreted in the text, but above all, they could be an allusion to various types of proto-humans – or precursors of current humanity – in various ways. different groups. What is clear in all the narratives is that the human body (homo sapien sapiens) as such was the last to appear, and things basically followed the same sequence, which is what we have continued throughout these chapters.

Furthermore, with the appearance of man, creation was concluded - according to the position of the monotheistic worldview - sealing the cycle of formation of the cosmos as a complete and perfect setting for human experience. If not, what can this apparent gap or change in that version mean? Enoch is the one who had given more details about the creation of things, and he is much earlier than the other narrators, so what would be the reason why the deity who spoke with Enoch told him that the animals had been produced in Yom Five, but only man was created on Yom Six - without any animal -? We can give interpretations, but the truth is that Enoch not only says this, but also adds the "process" and "patterns" that led to the appearance of the human: «And on the Sixth Day I ordered my Wisdom to create man from 6 bases: his meat from the Adamah; their blood from "the speed [of] the malachim and the densities-clouds their tendons-ligaments; and his hair from the grass of the Adamah; his soul of my breath and my spirit." (2 Enoch 30:10; or 11:58 of the Hebrew scroll).

Could there be a relationship between the Jaiah, Remesh and Behemah with the prelude to the appearance of man in a biological sense? I am not talking about sporadic evolution, but about deliberate genetic guidance: Intelligent Design. Was it the Wisdom of the deity that was behind the creation of man? As also stated in the Nag Hammadi manuscripts, there appears to be a spiritual collective representing a celestial kingdom - called 'Hokma' or 'Jajmah' (wisdom, science, wisdom, cunning) - which directed the operations of the Holy Spirit through of a possible civilization – or a group of them – of "angels" or superhuman beings that represent God and personify the Wisdom of these star systems. As Ra, Alex Collier, Zecharia Sitchin, Sixto Paz, and many others explain, this was an ambivalent process, with the intervention of more than 7 extraterrestrial civilizations (one version states that there were between 17 and 22 species) to produce human DNA. of the inhabitants of this planet, where the main participant – and who had the last word in genetic improvement – was the consciousness known as Iaheveh, the god of the Hebrews.

An example of what I mean regarding the existence of various human races (see RS2, chapters 2, 3 and 4) lies in the fact that these

Nefesh Jaiah (living souls) have an association with a Greek word: Aima. The similarity between the Greek sound Aima (blood) and the Spanish Alma, as well as the Hebrew Almah (maiden, virgin) is notable. If we remember Paul's phrase in Athens before the Areopagus, he stated that "God made the entire human race from one blood", which is similar to saying that from one lineage, but referring to the race of Adam. If we were to look for a hidden code in Alma like Aima, it would seem to us as if it said that from one soul he made the entire race. Nefesh is soul in the sense of being, of creature of essence, as when it was said to Noah: "the life of every soul is in its blood." The Jai of all Nefesh is in his Dam; the Zoi of all Psiji is in its Aima. The mystical meaning of this statement is that the life-existence-experience of the being depends on its soul-being, which is its race. On a level of the psyche, what happens to a creature is subject to the Collective Consciousness, and within it it is subject to the Planetary Consciousness, as well as the Racial Consciousness, and in it the Social Consciousness. Another verse says that "life is in the blood"; in the Dam is the Jai; in Aima there is Zoi.

Let us properly analyze the encryption encoded here: Dam is the blood or Aima (soul, essence) that gives life to a creature, and this comes from God, that is, from Adam (A-Dam). Blood is, therefore, the physical – and even psychic – manifestation of the integrity of the patterns of the divine being. The Greek word Aima itself is formed from the letters Alpha, Iota, Mi and Alpha, whose esoteric aspect is that the Celestial Principle (A) (I) in Matter (M) is God (A), projected on it. If the Dam (blood) comes from Adam (god), in it is Zoi (life), that is, Jevah. Eve is in Adam-God through the life that provides motor skills and genetic patterns to a body at the Corporal level. In the same way, in the soul there is Zoi, that is, the spark of the Holy Spirit that revitalizes our being, activates us and renews us, but above all these things, the spirit that revives us. If man becomes a "nefesh jaiah", or living being - whose metaphysical context is the capacity for spiritual transcendence - why is that same appreciation used BEFORE the human being "formed", to use it in the application of the life characteristic of apparent animals?

The Ape Man

The Oahspe says that the first terrestrial man lived in Ha'k (darkness), and was called Asu (the Asuans). Beings – which traditions have defined as "angels" (of 4th vibration density) – who had already lived in other worlds were invited to inhabit the Asu bodies. Being inexperienced in this, they lived with the common Asu and were "tempted" and "ate of the tree of life" and contemplated their own nakedness. There the first new Asu race was born, called 'Man' (or I'hin). These angels, due to the affliction they suffered, abandoned those bodies, and then Iaheveh came to free the Earth from Se'Mu affliction. From then on, those disembodied angels would be in charge of being "guardian angels" of their own people, that is, the human beings.

Among the fragments of an old manuscript attributed to the patriarch Abraham (dated to about 4,000 years old) there are some references that already at that time implied that human existence must develop throughout eternity in the infinite worlds that populate the cosmos, and that "angels" appear to be evolved beings who transcended the human form to which they once belonged (i.e., if angels dwell in the sky and inhabit the stars, it is possible that they would have been human inhabitants of those sidereal places): « The worlds are infinite and man has to live in all those that exist today; but creation continues and does not end. All worlds communicate with each other in love and justice, and my god is magnified in this. All the sons of my god, whom you call angels, were men..." (Text of the Secret Testament of Abraham). Just as with so much literature that already talked about this millennia ago, man seems to be a creation that exists throughout the universe, and seems to be a consciousness (soul) that came from a previous universe and/

or   a   spiritual   universe   (that

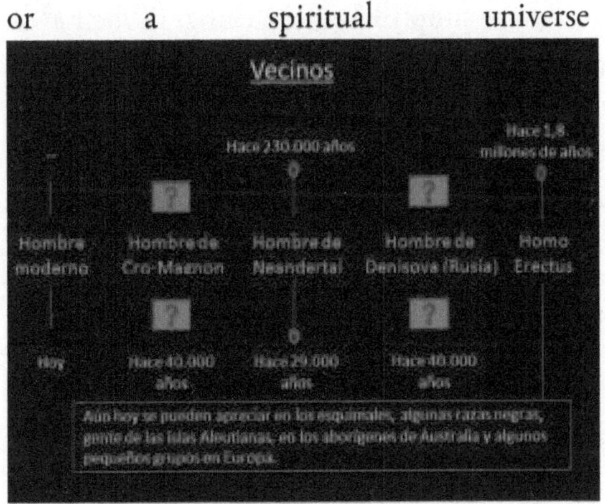

Shamaim') from verse 1 of Genesis 1.

In Amerindian cultures it is believed that there were several "ages" before the one that constituted the human era - and the previous one at the same time - and man already existed, but had been converted into an ape (the Ehecatonatiuh era of Aztec sources). This humanity would have been a prior design to the modern human body, and is described as "monkey" in the Celtic records, who also say – like other peoples – that this simian human race was almost entirely destroyed by the fall of an asteroid. According to ancient records, before or during the appearance of human beings there were giants in our world - according to the Aztec records - and they were in the First Age (that of Nahui-Océlotl), which was preceded by the Nahui-Ehécatl era, devastated by a "hurricane" or mega disaster that supposedly "turned" its survivors into monkeys. The Celtic Book of Creation Kolbrin records: «They tell us how the Selok monkey tribe, led by celestial men, perished by flames before the Valley of Lod; Only one monkey reached the upper heights [of the] cave. When the celestial man was reborn from the monkey in the cave of Affliction, he could taste the fruits of the earth, and drink its waters, and feel the freshness of its winds [...] Man, created from earthly substance alone, "He could not simply know the things of the Earth, nor could he by himself subject [to the] Spirit."

The Celtic story – which seems to have a certain symbolic resemblance to the story of Sodom and Gomorrah – states that this destruction was caused, basically, by an asteroid (and in other parts of the Kolbrin other similar events are cited that occurred in our prehistory destroying everything under the same description of asteroids that fell to Earth), and coincides with countless remote accounts from many cultures, including the Aztec itself, as it maintains that the following era – in which mostly divine or immortal beings lived on this world -, Nahui-Quiahuitl, was devastated by fire from heaven (asteroids or meteor shower). Why was man not created directly? According to the Popol-Vuh (Mayan records), an attempt was made to create man for the first time, but there was no success, but the subsequent trial did work. This highlights a large part of other stories, implying that the "gods" themselves who worked on genetics and life, although they knew what they wanted to do – since the human seemed to be a prototype already existing in the universe – did not know how. to do so, and it took them some time to find their ideal human, doing different biological and embryonic tests with the species that seemed to have the most similarities, at least on Earth: primates.

The Nag Hammadi texts, especially - as well as the Vedic records of India - declare that man already existed before coming to this world, but lived in other worlds and in other states of "vibration", "consciousness" or "dimensionality". ", so he needed a body as close as possible to the benefits of this level to experience this "physicality" - which the already existing forms of life (those of Second Density) did not provide -; Therefore, the "gods" sought to create a biological organism that was as competent as possible to be a vehicle for this immortal soul (man) that wanted to come from these other worlds. The Sumerian novel 'Atra-Hasis' tells that superhuman beings worked hard on earth, "Their task was considerable, Their work heavy, their labor infinite." These were of a race called 'Anunnaki' (which the Tanak calls "the children of Anak"), but they belonged to a subgroup called 'Igigi' - or 'igigu' - who obeyed the orders of the superior Anunnaki headed by the heir Enlil. : "[And these gods] [...] [The Igigu] (had) to excavate [the watercourses], [And open the canals] that give life to the earth. [Thus, they opened] the course of the Tigris, [And afterwards, [that of the Euphrates]."

Their arduous work exhausted them, "[For a thousand [...] years] they devoted themselves to the task – [After having accumulated [...]] all the mountains, [They counted the years] worked. [After having organized [...]] the great southern swamp, [They counted [the years] worked, [(For) two thousand five hundred years, and more, They had, day and night, Endured [this heavy load! It is said that they, tired of this situation, began to complain and criticize their leaders, conspiring and ending up carrying out an attempted coup d'état, attacking the seat of power at that time. It was 'Zu' (the great sage) who infiltrated among them seemed to motivate this, according to Hittite records (which adds that Zu tried to make a throne for himself on high and be like the Almighty), and curiously in the 'Treaty on the Origin of the World ', from Nag Hammadi, in the same order of events, the 'Instructor' appears, through whom the model is designed to create man on this sphere (I treated the entire explanation on this matter in detail in RS2, from page. 198).

They seized the mission control center and surrounded Enlil's mansion: "[Your pa]lace is sur[rounded], my Lord! The combat[has]extended to your doorstep! – Your palace is surrounded, oh Enlil! [...] Enlil ordered that the weapons be brought to his house, - Then he opened his mouth And addressed Nuska, his page: "Nuska, build a barricade before your door! - Take your weapons and put yourself at my command!» Faced with this threat, they recommended to Enlil to call his father Anu, who had decided to remain in orbit, or somewhere outside the Earth, in a celestial abode, and "Anu, the king of [Heaven], presided (the meeting), And the king of the Apsu, Enki, listened to [everything (?)], While the great Anun[naku] [sat] down, Enlil stood up: the debate [was] open[ed]. This urgent assembly was organized to establish the guidelines that were going to be carried out, since the Igigu refused to continue carrying this work. They gave their complaints, while Enki, Enlil's brother, supported them and recommended a subterfuge: "But there is [a remedy for this situation (?)]: Since [Belet-ili, the Matrix], is here, Let it make a prot[otype of man]: He will be the one who bears the yoke [of the gods (?)]- [Whoever bears the [y]oke [of the Igigu (?)]: [Will be the Man who loads] with his [work]!»

Belet-ili – or 'Mummy' - assures that he would do it with the help of Enki, since he was an expert in these subjects - which we could call "genetic" -, and he affirms that then a god must be sacrificed, and his blood It had to be mixed with clay to make a human design. These independent stories relate the combination of the gods' own essence with a pre-designed or existing model, to produce the human who was to be created: "The archons met in assembly and said: "Come, let us take land and create a man of mud". And they modeled their creature, making it completely of earth. (The Reality of the Powers 1:6-7, Nag Hammadi). In another treatise from this library it is said that the authorities "ejaculated their sperm" in the "navel of the Earth" to create life, and it gives the guidelines of how this development was carried out until reaching homo sapiens sapiens. Since there would be too many sources to cite and too much to write, I will try to summarize as best as possible all these points - most of which I already covered in RS2 - adding here that everything clearly points to genetic engineering being used to create the body. biological of the human being - speaking even of the use of the combined "elements of the earth" for the structural manufacture of a transport (body, avatar), but that in the first instance did not work, although theoretically everything was fine -.

The body seemed "alive", but it lacked a soul (crew), which they explained as a state of trance, coma or 'Tardema' (deep sleep). These attempts were not satisfactory from the beginning, and there were different attempts to create the "ideal" man, but when they achieved it they saw a danger and "mortalized" him again. Regarding the first process of said manufacturing, Henoch wrote: «And I gave it 7 qualities, the 7 for the flesh: sight for the eyes; the smell for the soul; the joy for the ligaments-tendons; taste for blood, patience-tolerance for bones; meekness-tranquility for thought. And I thought why I said-I called the Word cunning-wise because from the existing that is not seen and seen I made man, both dead and living, and the image-figure knows-knows Word and there is none in all Creation like it: small in greatness and great in smallness. And I placed him as the second guardian angel on the Earth, honored-upright and great, and respected-venerated. And I set him king of the Earth and there was no one like him among my wise-astute ones. And there was no equal to him on the Earth of all my creations.

And he continues writing: «And I named him the 4 winds: from the East, from the West, from the North [and] from the South. And I put him in custody-appointed 4 stars of the cooled-dry ones and called his name Adam. And I showed the wills-dispositions and he looked at the two paths: light and darkness, and I told them: That is going well and that is bad to direct, [I must] know if he has cunning-discernment towards me with hatred towards the direction excellent, [and who] of his seed loves me. And I saw my project but he did not know about the project, and if he knows he will do it wrong, to sin and everything of his will be sin, and I said: after sin there is no Word but death. [...] And I took the last letter of his name and called his name, they were him: Adam, [and] they [are] humanity and the living." (2nd Enoch 30:11-16) Cold stars? This was a direct way of saying "planet." Originally man in our solar system lived on 4 planets? Did he name it the four winds? The acronyms of the four addresses in ancient Greek were 'A', 'D', 'A' and 'M'. And, what is that letter that he took for his partner's name? Javah (Eve) was defined as "am kal-jai" (mother of all life), so her name Javah derives from Jai (life), and means "giver of life", and also, the first and last letter of the three that make up the word 'ADM' (Adam, man) form 'A.M' (mother).

It was logical that she would be a mother if she were a woman, because even the rest of the creatures procreated and gave birth to their fellow men, so what this allusion intends to show is that Javah would give birth to stately beings, it would be a womb for the conception of a glorious race that embodies souls from the heavenly abodes. Knowing that this would happen was why "Ialdabaot told the authorities who were with him: "Come, let us create a human being in the image of God and with the likeness of ourselves, so that this image may give us light." (Secret Book of John 9:1) Ialdabaot knew that the lordly man had a spark superior to him and wanted to possess it, but upon seeing his true potential he repented and destroyed this initial mold (body) producing a new mortal one, and instead to dominate everything, was dominated by the authorities to serve them and be so busy with trades and tasks that he never woke up to understand his true nature. This entire pattern was created in the image of the beings of the higher realms that Ialdabaot and his ilk had once beheld: ' they created with their powers and copied the features that had appeared. Each of the authorities contributed a psychic trait corresponding to the figure in the image they had seen. In the version of Genesis, all this must be deduced based on meticulous analysis, starting with understanding that the first created human was made in the "image and likeness" of the "deity", with the purpose of ruling over all things, which is essentially a call to self-control.

The Male-Female Projection

The story of Moses tells us that once all of the above has been carried out, and the stage has been prepared, he speaks in plural, in the second person, to determine that the man (that is, the human) is made: "ve-iamer elohim naaseh adam ba.tzelmenu k.dmutenu", which means "let us make Adam in our image like us." If Elohim is a god, and here he speaks of "let us do," certainly the deity is more than one. That being already has its own name (adam) so it was previously known what it was. However, these deities determine to do so on the material plane, and the model they rely on for this is to configure their body prototype based on their appearance, which would consist of it being similar to them. It is logical, all the time, that Adam and Elohim are the same thing, and what is below is the "image" of what is above, its living resemblance. And these gods add that said Adam dominates over the Deget of the sea, the Aof of the sky, over the Behemah and above all the Aretz, and over the reptiles that crawl on the earth (see RS1, page 166). This is not the story of Tarzan, Mowgli, Manimal, Aquaman or Ace Ventura. In reality, animals represent totems, which are personality characteristics. To dominate over animals is to dominate one's own instincts.

Verse 26 of chapter 1 of the Genesis of Moses says "ve.ibra elohim et-ha.adam ba.tzlemó, ba.tzlem elohim bará ató zakar ve.nekebah bará atam", which translated would be that the group of gods created Adam in his own image, in the image of the gods he created them, male and female he created them. Doesn't that seem a bit redundant? The Elohim is complete, with the masculine and feminine virtues of light, so that if he creates a being in his image, that being must have integrated the double component, masculine-feminine. By saying "he created them", it is not that he created two, but that the created Adam was not an individual but a collective, an entire race. Curiously, where it reads "et ha.adam ba.tlemó", the ET form is seen again, from the letters Alef and Tau, which represent the beginning and the end. Because? Because everything will be in the end as it was in the beginning. Masculine and feminine in one. And also the initials of these three words form AHB (Ahab), a word that means 'love'.

Apart from this, there is a set of initial letters Alef and Beit one after the other 3 times - like the 3 humanities -, the last one separated by the initials Zain (identifying the woman) and Vav (identifying the man). The mystery in the alphabetical code of Ahab (love) is the mystery of true marriage: there was a separation-death because the Woman did not join the Man while they were in Paradise, and that disunion, or cosmos, must be redeemed by middle of the meeting. That was its beginning and that will be its end, and that is the process of the role of others (ideal help) as a mirror of our own subconscious and unconscious distortions. That is the mystery of Love, which by extension is the mystery of Forgiveness. If a God had created man in his own image and likeness, and that image and likeness are male and female, then the original God's image has a male and a female (correctly spelled: male and female, or male and female). How could it be any other way? It would be absurd. You can't say you're making a statue in your image if it doesn't look like you. Then it's not your spitting image. It can only be your image if it is similar, because that is why it is called "image", as until today a statue is called an image in the Hebrew language, since it is identical to what it intends to emphasize or imitate.

If the created human is identical in image to his creator, and his creator makes that image "male and female", what, then, is the image of his maker? Voila! It literally says there that the image of man that he made was "male and female", and that modeling was based on the image of his Creator. Therefore, the Creator has within himself these two forms already incorporated into his appearance-character. Is androgynous the same as hermaphrodite? What is being a hermaphrodite? The myth of hermaphoditus is a Greek story that refers to wholeness, where there is no lack or deficiency. Therefore, God is neither male nor female, but everything in itself. If God were masculine, he would be dual and insane and macabre, because he would have placed man in sexual weakness, and that would make God the accomplice and first culprit of all world and historical problems of a sexual nature, from abortion, to rape, pedophilia, homosexuality, prostitution, sexual fever, pornography, transsexuality, bisexuality, infidelities or adulteries and other fornications. If you create something vulnerable and susceptible, the minimum is not to blame it later for falling because of the weakness you imposed on it.

God is total, non-dual, he is everything in himself. Being called masculine out of habit does not make it masculine. Simply put, in Hebrew there is no way of referring to it that is neutral, because in Hebrew there are no such possibilities as there are in other languages, and furthermore the language itself is dual. A very simple example, in Spanish you can say to a woman "tú", and to a man, "tú", but in Hebrew everything is dual: to a woman you say "at" (feminine you) and to a man "atá " (you masculine), and only in a few languages are there neutral forms. Let us also not forget that we have lived on a sexist planet for thousands of years, and any woman who, in most cases, would have been honest, would have been silenced, ridiculed or stoned, except for specific, exceptional and specific cases that occurred, such as in the Esther's story (given the conditions). If you wanted to get rid of a woman you just had to convince her to become a leader or a teacher.

Those towns only wanted to see a strong deity in the macho figure that they longed for. For this reason, almost no one understood why Yeshua had more female followers than followers and gave so much importance to the role of women, which also caused many debates among those close to him. Yeshua was not dual nor did he believe in duality, therefore, he never gave hints of machismo, unless he was misinterpreted (because when he said 'Father' he did not refer to a man or a male, but to the one who is All-Father). Machismo has been a great plague of the last 2 or 3 zodiac eras, to the point that Yaheveh himself could only be taken seriously by presenting himself as masculine. If the feminine had not existed before Jevah, it would not have made sense to say "it formed an Aishah", since it says nothing of their contextual and conceptual, morphogenetic, anatomical, hormonal or biological differences. If the woman complements the man, that is because at some point the man is not complete. Does God make mistakes? Do you hide your flaws by later creating fixes? The man was perfect, but not the man in the masculine sense, but in the sense of Adam, whose meaning is Christ.

In the existent and all encompassable there is not only matter and spirit: there is ether and there is psiji. What in the physical - or Asiah world - is sexual according to the organs of mammalian animals, is not so in the psiji (of the world of ideas, as Plato called it), neither in the spiritual, nor in the ethereal - or eternal -, which is the world Atzilut. The ideas of masculine and feminine exist since the beginning began, although not before the beginning of the beginning, because before there was time there was eternity, and before everything began there was already a One - who was and is completely complete -. The character projections of copulation only occur in the world Asiah, or what Genesis 1 calls 'Aretz' - which is matter -. At higher levels it is not created with the organs of a physical body. At the ether level it is created by thought; at the level of the spirit it is created with the word (Logos); at the level of the mind it is created with emotion; while at the level of matter it is created with action, copulating or manufacturing with the hands. In the Atzilut the ONE is projected by non-differentiated complementary emanations, except in its name, since its name identifies its masculine and feminine virtues.

Since they do not have sex or organs, they are neither male nor female beings, since there are no forms or images either. They are perfect seeing their reflection in front of them, because the two are one and the same part of the Totality. The first design was androgynous: "male and female he created them" (verse 27). However, this vegan man (verse 29) belonged to Yom Six, long before modern man, and Moses also recorded this when he wrote: "So the heavens and the earth were finished, and all the host of them. And God finished the work that he did on the seventh day; and he rested on the seventh day from all the work he did. (Gen. 2:1-2, RVA 60) What does this mean? That when the time for cessation of this project arrived, a period - Seventh - was imposed so that things would pass by themselves, like the seed that has already been sown and watered, now only expected to germinate and produce over time. Notice that he mentions a certain "army" or host, but at what point did he give details about the characteristics of said army of creation? It refers to all the consciousnesses that operate at multiple levels of density, and that direct the mechanisms of the lower levels. The conception of male spirits and female spirits was also projected within these vibration scales.

Androgynous means male-female, from the Greek words Andros (male) and Ginos (female) - but not in some type of transsexual aspect or sense - but as a complete being in itself. This sounds shocking at first, and some believe it is some kind of joke about sexual disorders, but the androgynous being is balanced, and is essentially devoid of the deficiencies of sexual need. There is nothing scandalous about this. And as religious people would say, where does the Bible say it? The Bible refers to totality as Perfect and complete. But even if this were a biological parameter, it does not conflict with anyone's religious principles either. Religion has persecuted those it considers impure, infidels, sinners, unconverted, apostates, lukewarm, inconstant, etc., and has determined with its judgments and indications who deserves heaven and who does not. To do this, they find passages that not even they understand, but they serve as a pretext to raise the accusing finger and condemn, violating – by this same act of lack of love – the commandments given by Yeshua: "if you understood what it means, I want mercy and not sacrifices, "You would not condemn the innocent."

Who are the innocent? WE ALL ARE. The ego wants us to believe that there are sinners and culprits. Revelation says that effeminates, sorcerers, dogs and many others who practice evil will not enter the holy city, but we are not speaking literally. The holy city is the state of completeness, the state of true Peace of God. By saying that certain people with their actions will not enter, he means that in the state of plenitude there will be no polarities or duality or distortions: there will be no imbalance in being. God does not reject any of his children. The Holy Spirit is responsible for correcting the Mind to eliminate from it all the diffusions of separatism and individuality typical of the ego. For this reason, Revelation also maintains that fornicators, adulterers, effeminate people, etc. they will go to the lake of sulfur and fire where the ego and its idea of separation would already be, because what it really wants to refer to is that all experiences are purified by karma and the awakening of individual, racial, collective, social consciousness and planetary. The fire purifies everything, and that fire is the tests and experiences of life, which is already hell for each one in his state of conviction that we are supposedly separated from God.

They bear fruit and multiply it

Continuing with verse 28 of chapter 1 of the book of Moses' Genesis, we find a mention of "bearing fruit and being fruitful": "prú ve.rabú". Prú from the form Pri (fruit), and Rabú from Arbé (a lot) and Rab (large). Another similar case appears in the synoptic gospels, when Yeshua affirms: "I have chosen you so that you bear much fruit, and that your fruit lasts." Could it be that the Master was suggesting to them that part of their ministry was to get married and have many children? The mystery of the words "be fruitful and multiply" lies in the understanding of the highest inductive level parameter. As Saul of Tarsus said, "if you have risen with Christ, set your sights on things above, not on things on earth", you have to understand the truth from above, not from the world, unless it has not yet been " risen" (awakened consciousness). "Be fruitful" was said to the soul, not to the body; That soul, which knew where it came from and why, had to "bear fruit" in what it was destined to do, and it knew it, since it received that letter and role – or Dharma – in the Entrevidas state, before incarnating.

We must make our soul fruitful and multiply. Otherwise, why should they copulate? That command came long before Jehovah-Mother came out of Adam-Father, since the soulful came first and then the corporeal. Multiplying is explained in the unnamed treatise on the origin of the world, from Nag Hammadi, where it is clear that the name Israel originally comes from the way of collectively identifying the Setite souls that would incarnate in this world. They, as a whole, were already called 'Church', from the Greek language, where Ekklesia means assembly, tumult or congregation, which is synonymous with Synagoge, or 'Synagogue'. He then adds that we must subjugate and dominate these animal life forms: "kabshah ve.radú". That? Own domain. It identifies self-control, but not in the way Christians perceive it, but in the realm of the Mind. It is about not giving room to animal thoughts. There are the "higher thoughts" and the lower thoughts, or those of the beasts (which are the principles of separation); the elevated thoughts, for their part, are those of the true Adam. The animal groups also symbolize the scales of consciousness of the sub-octaves of Second Density, which in turn is the same analogy of the awakening of consciousness in the Mind, its path of transcendence.

When talking about the transcendence of being - which religion calls Resurrection, due to the Spanish language influenced by Latin - it refers to the fact that there is separation in the world through a symbol of a greater separation that existed at the beginning (that is why it says "separate light from darkness", and then separating "waters from waters", since at a certain moment duality was projected in what until then had only been total and complete unity). Marriage symbolizes the path back to union with God. At first they were both in one, but then they separated, and since then they have been looking for a way to unite again, and after that symbol of union comes that of the family, which is followed by friends, neighbors, and fellow students. job; then consciousness increases towards a state in which it encompasses a neighborhood, a community and a town, and then reaches what is already a social consciousness, it passes to a larger and collective consciousness until reaching the state of global, racial and planetary consciousness., eventually galactic, to finally be a universal consciousness. That is the so-called 'Mind of Christ'. Only when they truly come together will they be one, and then they will know that they are in God and that they are God, that is, Christ or Adam. Marriage is not intended to multiply into more separations, as this would be duality.

Rather, what it implies is that true marriage - which is union with God - multiplies the fruits of God, which draw all others into union through the work of the Holy Spirit. It is the same thing that Yeshua said with other words, such as "I have called you to bear much fruit", but he did not mean that he had called them to procreate many children, but rather that they procreate more results of true love and true forgiveness, this is, of affiliation. The true children that are procreated - the fruits - do not come out of the union but come towards the union; they are not called from the inside out like the children of flesh and bones, but they are called from the outside in, like the children of the spirit. A clear example of the spiritual action and Right Mind that Yaheveh wanted to encrypt in these ideas that his angels taught Moses are outlined in the following verse of Genesis 1, where 29 says: "va.iamer elohim hanah natati lajem et- col-esheb, zera zera asher al-pnei col-ha.Aretz, ve.et-col-ha.etz asher-bo prí-etz zera zera, lachem ihih leachláh".

I could summarize this verse by arguing that the Collective Consciousness said that it had given to the projections of its Mind of all grass, reproductive codes before all the creation of illusion in which illusions, forms, ideas and concepts could be projected, and that would be the that I would experience. When saying 'Lachem' (to you) it is a play on the word 'Leajláh' (to eat), just like 'Lakaj' (to catch, take). In all three cases the common denominator aspect is an allusion to the experience that they are going to live, what they are going to learn and know, and what they are going to be part of. This passage that seems to say that humanity was ordered a strict and clear vegan diet, is, above all, an exhortation to the Holy Mind, the mind oblivious to judgments, which does not practice duality, which understands uniqueness and sonship, which includes love, piety and mercy. All these symbols are essential for the Mind, because whoever does not understand something as simple as respecting the life experience of a creature - however small it may be - will not understand much larger parameters, because whoever is not "pistio" in the little, much less it will be at most.

What is "pistio"? It is not "faithful" with respect to fidelity to administer something to someone, but with respect to faith, that is, love. If you do not have vision in the small, "how much less will you understand the heavenly?" And if you do not understand love in the smallest part of the dream, much less will you do so in the perception of the greatest, primary and primordial: you will not understand the source or its truth and reason. Elsewhere Yeshua says the same thing when he states: "let the little ones come to me," and adds that "theirs is the kingdom." It is not only an accurate allusion to innocence, nobility, simplicity of heart and humility, but also not to underestimate the smallest details and aspects of the symbols of the projection, since such are an example of the superior. If you understand the "smaller" parameters, you will understand the larger, broader and deeper ones. Ergo, that humanity did not eat meat or any product that came from an animal, why? Because the call was to a Right Mind, alien to irrationality, lightness, arbitrariness, the egoic mind. It was later said that "life is in the blood" and that "the life of every being is in its blood", as well as with respect to the soul and the flesh, since the body is a structure.

What I mean by this? We say "the body of the police", to refer to an entire body of consciences that form a totality. It is the same as saying "they are one flesh". As long as they are one, they are a complete and perfect identity. Thus, each complete and integral animal plays a role and has a meaning, both in the ecosystem and in the psychic perception of those who are close to it. Whoever destroys a body, destroys its value, and tells his unconscious that no symbol has value. This is why the Israelites were allowed to continue making sacrifices, but they misunderstood this and led to permanent animal genocide. Killing the Animal is an allegory about undoing the multiple characteristics of the ego, but in a mental sense, not a literal one. That is why it was said, "I want mercy and not sacrifices." The Mind, on the other hand, loses sensitivity by seeing murder as natural, which is a programming that also desensitizes it towards its neighbor, whose symbol is some type of animal character. We see our fellow humans as beasts, and therefore there has been no awareness of love and forgiveness for thousands of years when murdering animals for food, because the egoic Mind feels that when another dies for its own unconscious faults, it exempts itself from its own sacrifice, which is getting rid of the ego.

For this reason, those who still eat meat for "taste" feel it "tasty", because their unconscious mind enjoys punishing others, whom it considers guilty, and whom it judges, since it understands atonement only through suffering. On the other hand, he mostly does not feel the same when seeing the suffering of another human, because he sees himself reflected in him. That is why the Mind feels that an animal is worthy of dying, but not a human. The idea of error is such that the man himself who lives in the illusion of the ego believes that it is necessary that there be trials, suffering and punishment for atonement, and the guilt disappears, but it does not disappear by killing fellow human beings or animals. Eating meat symbolizes for the Unconscious Mind being a participant in the illusion, believing in it, loving it, enjoying its illusions and becoming attached to its figures. If we look carefully at the penultimate verse that ends chapter 1 of Genesis, it tells us that unlike humans, I give other creatures vegetables so that they could feed on them. Don't you think it's strange that he never presented the idea of carnivores? It is as if – taking it literally – it tried to assume that no animal ate another to survive or feed itself, but rather that they consumed "green grass."

More than two decades ago my father – Rabbi Felix G. Van Katch – had interpreted this as "knowledge" in a very simple or basic realm. Certainly, even if we went back to the time of the dinosaurs, half of them were carnivores. So green grass means something else, rather than a literal nutrition point of view. However, verse 30 uses the words 'Ierek Esheb' (green grass). The three-word phrase 'Ierek Esheb Leachlah' forms with its initials the voice 'Ial', which is "to serve" or be of use, as well as Iaal, which is a goat, while mathematically it gives 139, which is 'Gan Elohim' (garden of gods). The word Ierek (from Iarok) gives 310, like Arom (naked), a temura of Ierek being the form Rakei (empty); the word Esheb gives 372, like Sheba (seven), and is even its same color; the expression 'Ve.Ihi Ken' (and it was so) gives 101, as the form Melujah (kingdom). For its part, the form Leajlah is permuted as Kol-Elah, which in Aramaic is "all the gods," and in Hebrew it can be "all lamentation," "all oaths," or "all these." If we take all these definitions and put them together in the context, it does not speak to us about animals. On the contrary, it denotes a mortal human situation. In other words, it was before the human race to acquire knowledge of the truth or limit itself to knowing the experiences of the world without any understanding of the process of understanding and transcendence.

God took a Sabbath

The next chapter begins by saying: "ve.ikelu ha.shamaim ve.ha.aretz ve.kol-tzbaam", that is, "and completed the Shamaim and the Aretz and all their army." That last phrase "Ve.Ha.Aretz Ve.Kol Tzabaam" forms with its initials the word 'Tzelem' (image), since everything in the world is a living reflection of the superior Aretz. Subsequently it is said that the Collective Consciousness finished its work in a seventh period and took time in said cycle as a sabbatical aeon. What does this mean? That the Mind was developing the projection in 6 cycles and in a 7th period it stopped projecting the dream momentarily. This is the same in reverse, where for 7 cycles of development of evolution of consciousness we must return to the Shabbat which is outside of this eon. Here also lies the meaning of the rests relative to multiples of 7 beats. Here the numbers speak for themselves. The current level of consciousness is the 3rd, or Shlishí, of the base SH-L-SH (Shalosh), where the awakening of consciousness is separated into feminine and masculine, and between them the Lamed of learning-teaching, but the awakening from the dream of the level of consciousness of this octave it arrives in the 6th cycle, 6 being the number of Adam. There we see the complete oneness of consciousness with the masculine and the feminine, as it is read from 6 in Hebrew: Shesh. This is Shin-Shin.

Unlike this, Shabbat is Shin-Beit-Tau, referring to the developing consciousness at the beginning of the world and its end. In other words, consciousness completes its integration of all experiences within the dream and prepares for the Eighth Day, which is outside of this aeon. The eight (8) is Shmone, of the Shemen (Shin-Mem-Nun) which is the anointing oil, of the olive juice, of Christification, of the Christic Mind in the Totality. Then the last cycle arrives, which is the one that returns to non-time and non-space, which is the infinity of 9. In itself, 8 is already a call to infinity, a scale of consciousness out of this current octave to to enter infinity, and for this reason the horizontal 8 is a symbol of eternity. But 9 is the complete transcendence to infinity and plenitude. For this reason the number is called Teisha, where Tau is the end, and Shin is consciousness, since there is no longer consciousness of the Mind but the Totality in the Unity of the Absolute. This one has the letters Shin-Tau-Ain, where the Ain (eye), from the same sound as Ein (without, there is, nothing), in the context of what it simply Is, that is, the Complete Invisible Spirit Total Eternal Infinite Only Perfect Virgin.

The Eighth Day

Why do I speak of "eighth day"? As a general rule, we speak of the number 7, due to the levels of development of the illusion. But the 8 and the 9 are missing – even the '0' -, since with 9 numbers one starts for the multiplicities. The 8 symbolizes the ogdoada, and the 9 the infinite. After Maia's training process there is a time of calm and inactivity until what has been designed produces results. That is Day 8. On a metaphorical level we are in Yom Shmone (eighth era), but the final one will be the ninth, which is that of plenitude. The number 8 is the number of the letter Jet, which symbolizes Life, as well as the Hebrew word Ahab (Love). It is in the cycle of Love that we find ourselves, learning all the facets and levels of Love, that is, the configuration of the idea of separation of Pistis. There is Love in her: in her intention. All projection has intrinsic Love, and the learning scales and their types are all aspects that have intrinsic Love. It is not love that man understands, but True Love, which is understanding that we are all innocent, and that this It is nothing more than a dream, that there are no enemies, no death, no dangers... that WE ARE ALREADY HOME, but we don't realize it.

Let's look at some other ironies about the interpretations given to the text without understanding the intrinsic truth behind it all. Defenders of biblical literalism – such as creationists – maintain that the "days" of Genesis are just days (of 12 or 24 hours). The Hebrew word Yom is translated as Day, but if we went through the languages that arrived in the Mediterranean, we would find the word Ion, and later Eon. From there comes the Greek sound Aion, or Aiona (in Hebrew, Aiom means 'today', or today, the day we are in), which is sometimes translated from the NT as 'century'. Already in Spanish there is a clear difference between a day and a century, so that it is clearly something more, yes, related to a period of time. Precisely in Genesis 2:4 he begins by telling us: "eleh toldot ha.shabaim ve ha.Aretz." This translated is "these are the generations of the heavens and the earth." If what he is talking about is the Toldot of the Shamaim and the Artez, we cannot assume that a generation develops in 6 days, that is, 144 hours. In the usual context, a generation is the period of time in which a total of beings are part of a line of succession before or after a reference individual. In other words, he is telling us that the story in question is addressing the cycles of generations of both all those heavens and all the earths or physical worlds.

For this reason it is almost the same as conceiving the interpretation of an astronomical generation - which consists of approximately 2,160 years - as a comparative example to elucidate cycles of a very long time in which said events occurred. Another detail here lies in the translation error, when it translates "when they were created" starting from the Hebrew expression of the text that reads, 'Be.Hi.Braam'. The particle 'Be' is "in", not "when". The text, however, says that what was previously narrated corresponds to the generations or cycles of heaven and earth "in which they were created." We then have that such heavens and earth-matter were created throughout 6 eras that are also understood as celestial generations. This is all reinforced when reading the following sentence, which adds, "ba-yom asot", that is, "on [the] day [that] he made them". Wait! Did you do them in 6 days or in one day? Yom refers to time, periods, eras, ages, cycles. The 24-hour cycles of Earth's rotation take their name from the idea of cycles that are understood in the heavens. Ergo, as Genesis 1:1 says, "Barashit (ONE) created Elohim (1), the beginning-end of the Heavens (2) and the beginning-end of Aretz (3), and this includes these "toldot".

In verse 5 of chapter 2 he tells us that Yaheveh had not rained on the Aretz, contrary to what would be "made it rain". This is because in essence the passage implies that Yaheveh had not yet descended to Earth. He then adds that Adam did not work the Adamah. The codes in this passage are too relevant to be overlooked. It introduces the words Terem, Shij and Ein for the first time. Terem is a "not yet", while Shij is a plant. That is to say, when that happened there were still no plants, and he adds that there was no Esheb either. This context corresponds to the same Sumerian reference to Creation by maintaining that at the beginning of this planet there were initially no plants or herbs. Strictly speaking, at the time of planetary formation, oceanic alkalinization and the regulation of atmospheric gases, biological life had not yet begun.

The text tries to explain that the plants had not grown because it had not yet started to rain, and this is quite logical, since rain is part of the hydrological cycle. Not only the orbits had to be adjusted, but the air and ocean currents and the movement of the tectonic layers, which would begin to occur as a result of the cosmic influences of solar radiation. However, this verse also gives the impression of arguing that this type of development of the vegetation had not occurred because apart from that man did not work the land. We find that it no longer speaks to us only of Aretz but of Adamah, which we can deduce at the level of form as a distinction between a solid planet and the part where humans live. Here we can see that it is implied that the vegetation depended on the beginning of the hydrological cycle and on humans cultivating the Earth, and previously it tells us that "it was given" seeds that would be its sustenance. Expressed in another way, life had already occurred on other levels, and the forms of its production had been projected, but the ability to replicate life was given to cosmic humanity, not simply in its species, but to replicate the forms biological, whether zoological or vegetable.

The most interesting code here is that last quote from verse 5: "Ve.adam ein la.abod et-ha.adamah". At the level of the Mind, Adam is Elohim, Ein is the previous Beginning, Abod is death, Et is the predestination of time, and Adamah is the Sethite kingdom. This means that in that cycle humanity had not yet arrived in this cosmos from the superior kingdoms; the Elohim was not yet dead-separated within the parameter of beginning and end of the collective Adamah – the time that Adamah would be in a state of death -. Verse 6 says "ve.ed iaaleh min-ha.aretz ve.ha.shkah et-kol-pnei-ha.adamah". The word Ed means "testimony" or "witness", that is, at the level of the Mind, the Collective Unconscious determined that the illusion of the Maia-Anicca dream be experienced in the midst of the physical worlds. That is why 'Ed Iaaleh' is "testimony that stands up" or that is recorded. And he adds that this testimony soaks or permeates the entire surface of the physical worlds where human life was configured to be projected. That is why it is differentiated between the Aretz worlds (merely physical) and the Adamah (suitable for the development of Third Density consciousness onwards).

Part V

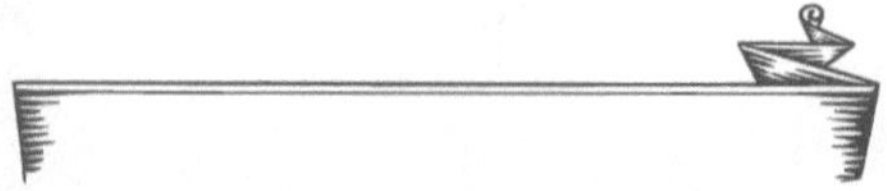

THE WRONG MIND

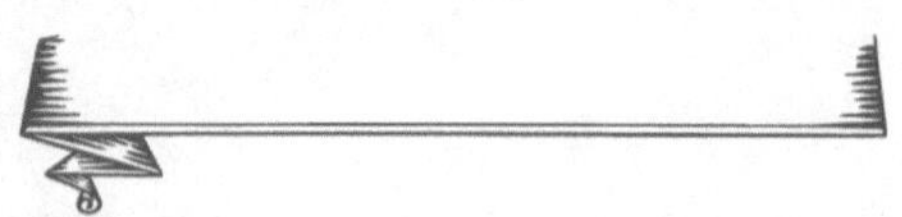

« Error cannot really threaten the truth, which can always resist
it. In reality, only error is vulnerable.
You are free to establish your kingdom wherever you see fit,
but you cannot but choose wisely if you remember this:
The spirit is eternally in a state of grace.
Your reality is only spirit.
Therefore, you are eternally in a state of grace."
(ACIM 3.V.1-6)

Cosmic Particles of Light

In verse 7 he explains: "ve.itzer yaheveh elohim et-ha.adam afar min-ha.adamah, ve.ipej ba.afin nshmat chaiim ve.ihi ha.adam la.nefesh jaiah." What is Afar? They translate it as "dust", but in reality it is "remnant" (see RS2, Why Was It Formed from Dust? - page 142), referring to "particles". The reference to the fact that humans were created material appears in various mythologies using expressions such as "dust", "earth", "clay", "mud" or even "corn". Why corn? In Amerindian myth, man was made of corn because at an archetypal level, corn was the basis of food. That is, an analogy with "bread", "fish" or "wheat" in Hebrew culture. In essence this refers to physical matter, but on a metaphysical level the dust is cosmic, it is the individualized particles of a single thing. Where does dust originate from? Of the earth, being carried by the wind because it is the superficial or lightest part of the soil. Those particles of soil fly with the wind and settle everywhere. That is the meaning of the man made Afar (propagation) in the Adamah worlds, replicated by all the physical worlds of the Maia-Anicca projection where biological life can occur.

The text can be read and translated as follows: "And Iaheveh produced Elohim the beginning and the end of Man dust from among the Adamah." Thus, "elohim et ha adam", means that Elohim is the Adam "propagated among the Adamah", or in other words, the humans who were dispersed throughout the worlds of this physical cosmos are the Elohim, within the parameter of time. ("et"). However, if you want to see God (Elohim) you will see him through all the Adam that populate the cosmos. The same biblical passage then maintains that the breath (neshmat) of lives (chaiim) was "blown" (ipej) into the nose (af), in such a way that the material human had the capacity to be in a state of soul (nefesh) with life (jaiah). Previously we had seen that life forms that received the qualification of Nefesh Jaiah had already been placed on the Aretz worlds, but now this same qualification was attributed to the human-god. We understand, then, that until now man did not have that virtue, which he received since the Nishmat was breathed into him that granted him lives, not life. What is the difference in terms? Life is existing, but "lives" is the multiplicity of those experiences. Chaiim is mistakenly believed to be "life," even though it is a plural, and is nevertheless used in other connotations such as "eternal life" (chaiim ad-olam), or more correctly translated as "lives in the worlds."

The same thing happens when the same idea is translated into the Greek language, whether to say eternal life or immortality – as far as the Bible is concerned. But if it is said that there is an existence, or life, which is prolonged, by extension it should be the qualifier that denotes the appreciation, not the subject. In other words, if it says "lives" it is because there are many lives, not a single long life. First of all, the word "eternal" does not exist in Hebrew; The word Olam is used, which means "world." Olam is written with Ain, Lamed and Mem (Alam), whose letter formation means that the One (Ain) connects (Lamed) with the world (Mem), adding another fact: Ain-Lamed refers to the High One, used as an article "on" or "above", that is, "what is above the world". Secondly, if the human existence and experience was given to him, just like other beings before him – which were interpreted to be simply "animals" – like many lives, it means that the beings that were before him did not have: 1st self-awareness, 2nd the ability to evolve. This coincides with the Material of Ra, explaining that the primary bodies that precede the Third Density (human as we understand it, or humanoid) were in Second Density, that is, without consciousness of themselves, however, incapable of aspiring to the transcendence, that is, the awakening of consciousness – and thus progress to higher forms of consciousness -.

With the notaricon kabalistic system we can take the initial letters that go from the sixth to the penultimate of this verse and form the expression: "ve.ha.am ben javah", which translated is, "and the people son of life", understanding by Javah the name Eve. Likewise, with fewer letters, this time final ones, we can form the phrase: "ha.jevatim shem", which means, "the so-called living ones", or "there are the living ones". Another fact is that the appreciation "le.nefesh" can be understood as descent (Naf) of the Shin, which symbolizes consciousness. This is – even in the context – that Adam begins to descend as consciousness through the experience of life, which we can define as experiencing within Maia's dream.

Metaphorical Jordan

In verse 8 we find that it says: "Ve.ita iaheveh elohim gan-ba.eden ma.kedem, ve.ishem sham et-ha.adam asher itzar", which translated would be that the deity Iaheveh planted a garden in Eden, at east, and placed there the man he created, but who and what are we talking about? The Gan (garden, orchard), whose symbol is the

Jordan, represents the immortal kingdoms of Adamah in the aeons of the Son. That is the Eden with which an earthly paradise in Mesopotamia was later named in the times of the first humans on this planet. The "Gan ba.Eden" identifies the "river of water of life" from another symbol introduced by the Holy Spirit in the vision known as Revelation. This is the return to paradise, to "the land flowing with milk and honey" which are the states of completeness that are achieved throughout the awakened experiences in the higher worlds. The 12 fruits of the Tree of Life are the 12 states-stays of Plenitude, projected in the illusion of dimensions as the 12 tones and their respective 12 overtones. For this reason, the text has the reference "ma.kedem", which is also "mi-kadima", that is, by which one advances. There man was originally placed, at the source of Sdom and Amrah in the kingdoms of Armozel, Oroiel and Daveithai.

This is essentially what we observe in verse 9 when it says that Elohim caused to grow in the Adamah every "tree pleasing to the eye" and also "good for eating." It defines two main trees in the narrative, since one is that of truth and the other is that of deception, that is, of duality. That is why he says that they were all pleasant to see, since no good god would place danger in a place of innocent beings, unless the situation was under control. A dream is under control, so there is no danger in the garden. These were two aspects of interest: walking in perfection or entering into duality. Therein lies the "good to see" appreciation, that is, positives on which to put focus and attention, on which to direct oneself. That is, it is pleasant, and it is "good to eat", that is, positive for learning, for developing, for being. Verse 10 tells us about the River, as in Revelation, which is the flow of light and truth of Fullness, and has 4 arms that "kiss" the garden (symbolized in the Jordan River). The kiss is a symbol of love and loyalty, and this is that the river is the direct connections of the 4 imperishable kingdoms starting from a common source which is the Totality derived from the Son.

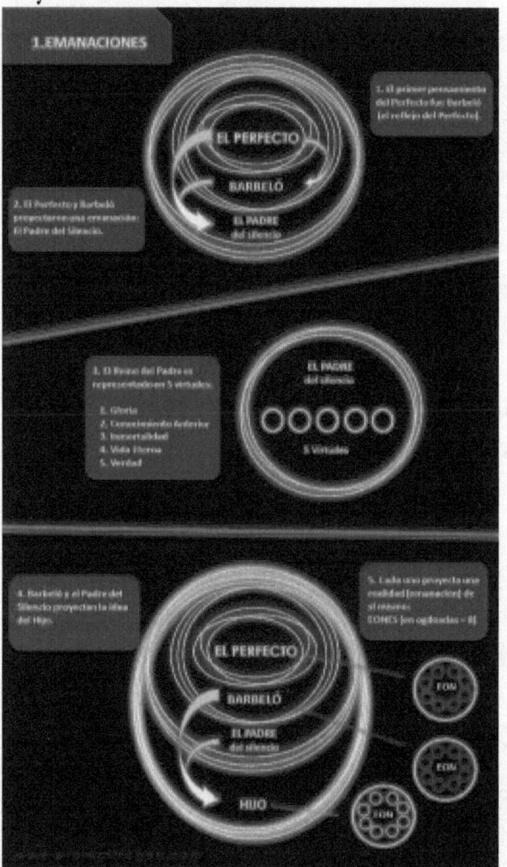

Just as the Hebrew says Neshek (kiss), it also says Rosh (head), not arms, when referring to the 4 sections of the flow of the water of Eternal Life. The 4 sections identify the 4 parts or conceptual sections of the brain, and are projected from the geometric perspective as the sphere of Silence and the sphere of the First Thought (Glory), which produce the vesica piscis. From within the Son emanates through a Cross that divides the central oval into 4 sections. These 4 sections can also be seen as the reflections or inverted aspects of the Left Hemisphere and Right Hemisphere of the brain which are defined as lobes. The front left part is the logical geometric, while the back right is the other logical geometric, that is, the masculine is on the left in front, and the right is on the right in the back. For its part, the left rear is experiential masculine, while the right front is experiential feminine. The masculine logical geometrician is triangle and square (3 angles and 4 angles) in two dimensions, or tetrahedron and cube in 3 dimensions; while the feminine logical geometric is a triangle and pentagon (3 angles and 5 angles) in two dimensions, or tetrahedron, icosahedron and dodecahedron in three dimensions.

The subsequent verses tell us about the names and some descriptions of the 4 rivers or capillaries of water that bathed the region. The first, Pishon, is an allusion to the realm of the Armozel star; the second, Gijón, is alluding to Oroiel; Jidekel is allusive to Daveithe; and Prat symbolizes Elelet. It is notorious that Prat (in Castilian Euphrates), was associated with the last of these, since the Euphrates symbolizes the division between the Right and the Left. The Euphrates River is located in Iraq, which was formerly Babylonia, and the extent was generally known as Mesopotamia. It is called "the cradle of civilization". Actually the name 'Babylonia' is Greek, as a formation of the original Chaldean Babili, and the Hebrew Babel. From Babel came the river that opened in four and emptied into the Persian Gulf and the Indian Ocean, or South Sea. Babel represents Barbelo; the southern sea are the universes; the "cradle of civilization" is, in effect, the cradle of every race in all that has been produced. Babel is the image of the first aeons, just as Egypt is the image of the cosmos – and between the two is Israel, which symbolizes the Setites -. Israel spent 400 years in Egypt, as it is a Baktun cycle (a short Dan), and it symbolizes the letter Tau, because that is where their servitude had to end. But he spent 30 more years, which were under servitude. 30 is Lamed, a symbol of learning and teaching.

The Israelites were liberated from Egypt, liberated from the Canaanites, liberated from the Philistines, carried away by the Assyrians, carried away by the Babylonians, liberated by the Persians, attacked by the Greeks, invaded and dispersed by the Romans, tortured and reduced by the Germans. harassed by the Palestinians. What do these 10 stages symbolize? The different lives of man with their own vicissitudes. These disagreements are only repeated in different cycles until the root of the problem is healed: the Mind. Thus, 70 years of Babylonian captivity was another stage of learning, where 70 is Ain, being observance, vigilance, attention, focusing and direction. Furthermore, the text of Genesis 2:10-14 tells us that there was gold, onyx and bdellium in Javilah (joy, rejoicing), these stones and minerals symbolizing the virtues of the kingdoms of the immortals. Javilah is the symbol for the abode of Adama, just as Cush is for the abode of the Great Set, or Ashur is for the Setite race. When this water irrigated these kingdoms, already bathed in its light, the Setite humanity came, as the text says: "he placed it in the Gan Eden", that is, in the region of the river of Eden, which is analogous to the Jordan (Yarden-Garden-Garden).

The Hebrew name Eden is formed from the letters Ain, Dalet and Nun, which numerically is 70-4-700 or 70-4-50. The figure 124 corresponds to the Greek word 'Anthropos' (man). Ain is the mystical way of referring to Adamas, just as the Aleph is used in the projection for the man who entered the dream. Ain symbolizes the eye (the Greek Omicron), as used in their iconography by both Ra and his apprentice Horus; Dalet and Nun symbolize the elemental cycles within the projection, but in essence they represent equilibrium and balance. Strictly speaking, E-Den represents the man who must know himself. If we read Ain with Dalet it is Ed, which is both a root of Edet (assembly, congregation) and itself the understanding of 'testimony'. This means learning and experimenting in order to learn and know. For its part, Nun is infinity. Thus, Ed-N is to bear witness to the infinite, that is, to know in order to testify; know-experience the cosmos to understand, be aware and know oneself.

The Immortal Mathematics

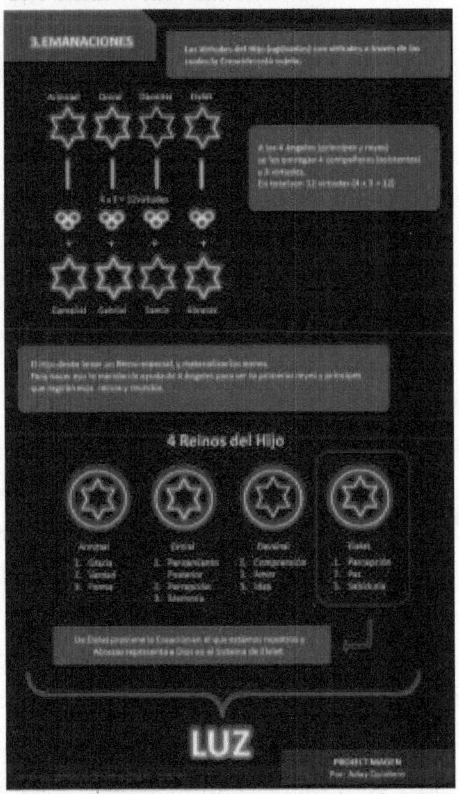

We can look at the mathematics of all this, apart from its geometry. Everything is in one as a cell, and just as conception occurs, the smaller cell (sperm) enters the larger one (egg). The egg is the Silence, the sperm is the Mind (the Glory, the First Thought), and upon entering the Silence the First Thought, life was conceived, that is, the Son. This is the same golden pattern: 1, 1, 2, 3. The One is first, then the Silence, and from its reflection comes the 2nd, and to the universe comes the third, but all within the Monad (Unity). It is the same as a bi-plane reality that becomes three-dimensional, that is, a volume or first form is projected, in what until then was only pure essence, without images or figures.

The full and absolute being remains in the center, and when another sphere of reality is created, there are already 6 realities. One central and 5 peripheral. This is so following the Fibonacci sequence or golden ratio. 1, 1, 2, 3, 5. We find here the beginning of numerical values and their mysticism. Every value associated with '0' is equivalent to Ein and Ein Sof; the value '1' is Ein Sof Aur. The values of '2' are the base from which the multiplicity starts, but '3' is already the development and the form. Once these three are configured, the emanations of the totality of the divine races and the aeonic universe (of the Ogdoads producing) begin, which is symbolized with the number 4.

It is curious to speak of Ein as "nothing" even though it is not empty in the context of absence of consciousness, but absence of any "thing." The Spanish definition 'Nothing' comes from Sanskrit, where it means "sound and vibration." Ein is the absence of objects, images, things, material, projections, illusions, figures, ideas, concepts, etc. It is simply what IS in itself as the essence of the absolute without alterations or noise. Consciousness becoming conscious produces waves in Nun (the infinity of space), and that wave oscillation produces a frequency, and from there comes sound, energy and light, and later matter. That is the so-called OM of Hinduism, or sound principle from which everything began. We have the examples of number 2 with polarities, not in the sense of necessity or dependence, but as two sides of the same coin. They are not a Ying Yang of duality, but the integration of the same thing being reflected in its component aspects. It is not the point here to talk about the Masculine and the Feminine (in that sense I recommend my book 'Naked Sex, the Origin of Sexuality'), but the basis of understanding, because that is where the 4 comes from.

The 4 letters of the tetragrammaton (IHVH) are another example of this - although in reality there are only 3 base letters, since one is repeated (H) -. A more basic example in the world is the brain division that I have previously taught you, which is the "2x2" sequence. There are two high principles in the world-cosmos, which are Fire and Air, and two lower principles: Earth (which is female) and Water (which is male). Here is the most illuminating example of the separation in 4 of the brain and its geometric aspects, since the elements also have a geometric principle. Fire is Tetrahedron (triangle), Earth is Cube (square), Water is Icosahedron (triangle) and Air is Octahedron (triangle and square). Just as the feet touch the earth, as symbols of Earth and Water, the hands are in the air, as symbols of Air and Fire. The mixture of water and earth produced Nature, and she received maturity from fire, and spirit from air. This is how the integration of the 4 principles and their result as Life occurs, that is, pattern 5. Nature produced bodies according to the Image of Adam. The Man, of Life and Light – who was – came to be with Soul and Mind. This is Man as Ether (5th element, but first of all), with the 4 potential principles of creation: Life and Light, Soul and Mind. Life became Soul, and Light became Mind. In other words, Life is Faith-Love, and the soul is Identity-Personification-Individualization, while Light is Consciousness and Intelligent Energy, with Mind within the projection being the analogy of Intellect and Reason.

The 4 has its image in the projection on the upper level as Angles of Heaven, and on the lower level as Angles of the Earth. Each culture had different ways of representing this idea, with 4 gods or 4 animals, or with 4 angels. From the Son came the concept of 4 to 5, but he identified 3, as the Vedanta worldview, which has the masculine Vishnu and the feminine Shiva and the balance Brahma. Thus 3 is the balancing principle and the unifying principle. Thus, it unites the two principles creating a potential for both, which is nothing other than the feedback of one's own uniqueness. It does this by producing the number 4, that is, all the universes: their spaces, dimensions, planes, realities, volumes, proportions, eons. That is why the 4 elements come from there, the 4 cardinals, the 4 seasons, the 4 colors of the base elements of the human body (red, black, yellow and blue) and their 4 corresponding minerals (iron, carbon, sulfur and copper).. I remember the example of 3rd Enoch, where it speaks of the spirits that surround the 8 wheels (ofanim and galgalim) around the throne of Adonai Tzabaot, which are: Strong Wind, Hurricane Wind, Storm Wind and Earthquake Wind. Logically this is not strictly literal (but I will not deviate further from the topic to go into details). There is also talk of 4 Jayot (living beings) around Adonai Tzabaot, etc.

A clear example of the projection with the number 4 is the letter Dalet, which is the "door" to access reality. Man entered there, through man, being an idea within man. That is, of Adam, that is, the man-god. For this reason the cosmos, represented in the number 4 – since it came as an image of the 4 kingdoms (image in turn of the divisions of the "brain" of the ONE) – is called (identifies) ADAM. We know this from the names of the 4 directions or cardinals in ancient Greek: **A** rktos (north), **D** osis (west), **A** natol (east) and **M** esiba (south). If you combine the initials of the 4 angles it gives you the name. In all this the base of the Cross is always present as the center of the 4 points. Another example can be seen when Iaheveh chose 4 ancient regions in the Middle East to make them sacramental: the Garden of Eden, Mount Sin (Sinai), Mount Tzion (or Mount Moriah) and the Mount of the East (Arab). Another example of waves is those that develop naturally in brain processes: Beta, Alpha, Theta and Delta. But we will talk about this particular matter later.

The Greeks said that 4 were the natures of the gods: creators, guardians, givers of life and inspirers. Being this image of the superior, 4 is multiplied by 3 to give 12, like the 12 tones of the musical harmonic series, arithmetic sequence. 4 princes with their 3 virtues making up 12. This is the example in the world of the Greek pantheon with Zeus, Poseidon and Ephaistos, as Creators; Hestia, Athena and Ares, as Guards; Demeter, Hera and Artemis, as Givers of Life; Hermes, Aphrodite and Apollo, as Inspirers. This is the square, the base of the cube, its 6 faces and its 8 corners and its 24 angles. The first 3 forms the volume of a tetrahedron, or figure with 4 corners, 4 faces and 12 angles. This is the kingdom of the immortals, when the race of glorious lights, or first angels, began to withdraw there. The tetrahedron preceded the cube, but they all come from the sphere, whose symbol is the Udyat, or eye of Horus. Adepts and philosophers believe that the triangle with an inner circle at its top, or pyramid with the eye is a symbol of "illuminati power", but this deception is still a ridiculous fantasy of Freemasonry. The triangle is the symbol of the true gods, and the eye at its top is Man. The eye is the ONE. The triangle is the Monad of the 3 corners: the 3 primordial aeons. For this reason, from the tetrahedron - which is LIGHT (whose archetype of the Fire element) - came the hexahedron (cube), or projection of the

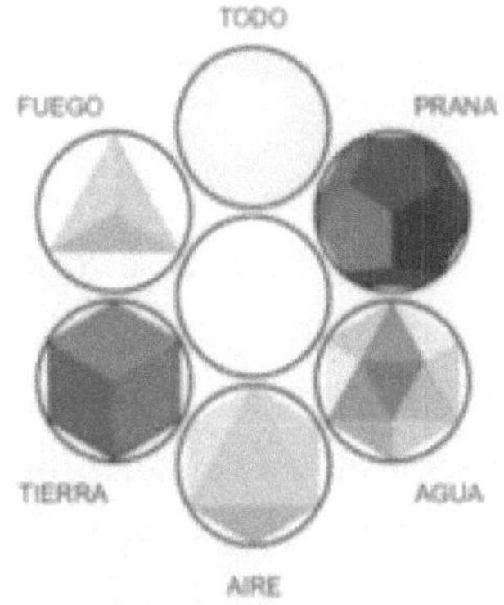

"earth". What land? The
Adamantine Earth, the kingdom of Adamas.

The beginning of these perceptible sounds were the vowels, which, as they appeared, so are the source of return to the origin. These are A, E, I, O, U. The first gods of illusion, the personifications of the ego, were precisely identified with these sounds, and in the same way they were used by the manifestations of light, because they are really the personification of the fountain. These can be associated with the Hebrew letters Aleph, Ain, Yud, He and Vav, as well as with the Greek letters Alpha, Eta, Iota, Omega and Upsilon - although Omicron can also be included as a substitute for Omega or equivalent of the Hebrew Ain -. The Oahspe defines this as "sounds produced by the wind", which would basically be 'E', 'O' and 'Ih'. So 'A' would be a variant of 'E', and 'U' a variant of 'O', which is explained with the use of the Hebrew letters Alef/Elef and Vav/Uau/Oau. From there came the original form of the tetragrammaton, and the IHV form as the root of it, which is a sound similar to 'Iao', but if it had been written Iud-Alef-Vav or Iud-Ain-Vav it would have had only two sounds (because the Aleph and the Ain complement the sound of a consonant, and if you put a vowel sound with the Aleph or the Ain, it would only reinforce that single sound). Consequently, the original basis of the name was IEO, written in Hebrew as Yud-He (Iah). The IEO form is intercalated with the Greek form IAO (Iota, Alpha and Omega) to identify the same as the Hebrew IHVH (see RS3, p. 34).

So that 3 sounds, such as three Ogdoadic spheres, project 5 forms (Hebrew vowels) and then 6 forms (Greek vowels) that end up constituting 7. Everything comes from sound, says the Vedanta tradition. The 3 ogdoados are 3 sound spheres of reality, that is, vibrations of consciousness as emanations. Each octave/octave has 7 light potentials: Red, Orange, Green, Cyan, Blue and Violet - as a spectrum of its light -, or sound potentials: Do, Re, Mi, Fa, Sol, La, Si. This is the invisible 24 of the Mystery of Mysteries. The ogdoads are withdrawn from their center, and for this reason the 3 spheres of the Monad (image in the projection of the corporeal, atmospheric and ethereal worlds) form a Tripod of Life that creates 7 realities, whose analogy is manifested in the 7 vowels. manifestas from the Greek: Alpha + Epsilon and Eta + Iota and Upsilon + Omicron and Omega. Here they constitute 4 groups, being 'E' analogous to 'H' – which is silent in Spanish and Hebrew – and Upsilon a combined form of 'I' with 'Y' with 'U'.

From this projection appears 5, which is 4+1. The 4 principles with a motor, just as it is seen that the Conscious Spirit does (analogy of the Ether) in the reality of the cosmos (the 4). We see this example with the 5 fingers of each hand and each foot: 4 times 5. This is the same as the elements (see RS3, p. 83), whether from the Chinese-Hindu perspective or from the Aztec or Greek: the Platonic solids. Each sound is linked to a geometric shape, a numerical value and a light pattern. From them emanates 6, which configures the final reality of the cosmos. Those are the 6 days. So, 1st was the Light, 2nd was the Firmament, 3rd the constitution of the Waters, 4th the Stars, 5th the Living, and 6th the Adam of the world. That Light is the ONE; that Firmament is the Sky of the Totality, the Infinite; the Waters are the emanations produced; the stars are the lights of the superior kingdoms; the living are the first races of the immortals; the 6th is the world, the cosmos, where the man formed from the projection came from. This is the number 6 – or the number of man – whose geometry is the hexagon. Just as the cube-hexahedron (the 6 faces) came from the tetrahedron, the octahedron came from it: two inverted pyramids with their points facing outwards. That is, the shape of the 8 outer faces, but with a hidden inner face. This means that the secret of 9 (eternity) is in 8.

The octahedron is the element Air, and just like the others, part of the triangle (the symbol of the Son, or symbol of the gods). The Air symbolizes the Spirit, the same one that in the 3rd Yom separated the waters, creating the divisions of the 3 consistencies: gaseous, liquid and solid. These three are an analogy of the 3 principles Ether, Atmos and Corpor. 5 is the golden ratio (Fi), and 6 is the Pi ratio. One is feminine and the other is masculine. The volume ratio of the octahedron also merges creating a star tetrahedron, whose bi-plane is known as a hexagram or "star of David." The star tetrahedron is the Merkabah, or chariot, the avatar of the soul, its dimensional vehicle. The first sphere is '0', the Ein, but the second one that intersects with it symbolizes the First Day or first process of creation; Then comes the 'Tripod of Life', accompanied by a 3rd, 4th and 5th cycle of formations, until reaching the 6th, which configures the 'Seed of Life'. This is the part of the 'Fruit of Life' and hence the 'Metatron Cube'. All these processes configure reality, called 'Flower of Life'.

By that great number 6, matter was produced, and from there the organic material, which starts from the element carbon, or C-12 (carbon 12), whose composition is the basis for biological life, essentially human life. This molecule is made up of 6 protons, 6 neutrons and 6 electrons. Yes, '666', which is not the number of the Beast, that is, of the ego. The ego was 6, and therefore man manifested in 6. These 6 principles of energy add up to 18 (6 without charge, 6 with positive charge and 6 with negative charge). If we take numerically the letters of the name ADAM in the Hebrew language (Alef-Dalet and Mem) they give us 18: A = 1; D = 4, M = 13. This is curious, just as if we take the value of the Mem in the Kabbalistic system (40) - because Aleph and Dalet are the same with this method -, this equation gives us 45. The human genetic composition It is mostly part of the number of combined base pairs of deoxyribonucleic acid (DNA). These chains are made up of 4 nucleic acids called Adenine, Guanine, Thymine and Cytosine. These 4 symbolize once again matter and materialized man. If we multiply the genes (45) by the acids, it gives us 180, which is half of the sphere (360). This symbolizes the Light (12 hours of the 180° of the sphere), among other things.

Someone may say that there is an error in my equation, since officially we have 46 chromosomes, not 45. Well, 46 in Greek numerals is ADAM (Alpha-Delta-Alpha-Mi), but whoever defines it with 45 is because the language Greek provides a symbolic and optional addition, because the mother provides only the 22 valuable chromosomes, and the father the other 22, and depending on the irradiation of consciousness, an additional chromosome is activated that accompanies the cell, be it XX or XY, simply to determine sexual gender, not biological structure. However, just like the 23 original books of Tanak, there are 22 elementary principles and one complementary one that is the determining factor. I will talk later about the rest of the numbers, because those related to the origin are 5, then the man came and then 7 and 8 were manifested. I will talk about the 7th later. Now the point is the state of the 8, which is the Adam in Gan Eden, in which Adama had 'Le.abdah' and also 'Le.shamerah'. Obed is work or service, but in the sense of death it is servitude and slavery, used as a reward for perdition and destruction. This means that what was a perspective of development was reversed "due to a misinterpretation", making Possibility an entry of the Mind into a dream of deception, called Mevet (death).

What he was supposed to "take care of" (Lishmor) turned against him, and he now had to "take care" of him. That "Obed" was the idea of separation, which brought the serpent: the ego. Religions received by transmission the idea of a demon whose archetype is the snake or dragon. Its symbolism comes from the sagacity of the ego, since the snake and the dragon have always been representations of cunning and aspects of knowledge. The representation of the coiled snake from the Muladhara identifies 7 blocks that the being has to understand its reality. These 7 have been defined in the mysticism of various cultures under the collective name of 'Death'. In the Hebrew language the designation of Nachash (serpent) was used to differentiate from Saraf in the sense of the mineral Najeshet (copper). This is not only because of the appearance of the copper snake or this color of the already molten metal, but of the powdered mineral, whose blue appearance has its own archetype. With the ego represented as a snake, blue is a connotation of elitism and the superiority complex of the Ich (I). Both the serpent-ego and the tree of good-evil experience were a possibility-opportunity in his hand, as is understood by Tanin (dragon) in temura: Noten (give). Man was free to decide, because that is what Free Will consists of. The opposite would be to control and manipulate without having the right to free opinion or choice.

Side as Shadow

Verse 16 maintains that at that point of everything being irrigated and the diamond breed being laid, they were allowed the possibility, but supposedly they were told not to test anything regarding good and evil. What's that? It says in Hebrew, "tob ve.raa." He is not talking about separate things but about the combination of two components. When saying "eat" or "don't eat", it is a play on words. Eating is Leajal, and being able to do is Leejol, which come from the phoneme Lej (go, go towards), which means that the Truth was in the consciousness so that the Mind could rationalize and judge whether or not it should experience what is duality.. That is why he adds that the "yom" that you eat from him "mot tamot". The yom that refers to the eon in which he will experience it, which would be of various levels of death. That is why it does not say "you will die", but "dead you will die". This refers to deaths upon deaths, deaths upon deaths, and varying levels of suffering. Once this was said, sin entered, for Elohim said "it is not good for Adam to be alone." By saying this we encounter the first idea of duality. Elohim, as Collective Consciousness, has the idea that being "single-minded" by Adam "was not good." Another way to translate Levad, in addition to "estar" is "ser", even as an idea of "referring".

For example, "it is not good for man to be referred to-identified-understood as silence". To be alone is to be in silence, it is only to exist what is. In other words, not just him, but something "else". Then it is said that all Life, that is, all life experience, is formed from Adamah. All of Adam's life experiences begin with processes in the "field" and finally processes in the "heaven" of the Firmament that "come to him", and he recognizes them. Giving a name is a reference to identify, to know the reason, the purpose and the destination of something, its objective (see RS1, p. 142). Man did not see himself reflected in any of these aspects of existence. In this case, Headquarters (field) are the physical worlds, and the birds reflect the beings of the projection skies, and in neither of them did man find a reflection of himself to learn from. The Headquarters is also the mystical way of referring to thoughts of darkness. However, the man did not identify himself with the wrong thoughts or with the elevated thoughts. He needed something that was on his level, something neuro. This means that Adam's soul came into existence in states that his unconscious chose as intermediate states of dual projection: not as a demon (Shed, or creature of the Shede, or Headquarters) nor as an angel (whose archetype is the bird).

The saying, "Woman was created for man" means that the Mind that created the universe did not emanate from itself, it came from the Christ. That is why Saul of Tarsus said in the Spirit that it was "because of," that is, because of another. The Adam that religion imagines in a human story is the representation of Christ, and neither is Christ lacking nor does it make sense that the man in the story that the terrestrials have not understood would have him. Even if it were literal, Adam, being created by that God, would not be dependent, lacking or needy. The true Adam - who is called Christ - is and was complete. If perfection were the creation that we perceive, why would that god create a human dependent on another human? What makes him sexually vulnerable? Why do you have someone to depend on? No. The Mother separated from God, that is, the Woman separated from the Man, that is, Life separated from the Christ. Genesis says that Adam lived 930 years, but this is a code: 900+30. The 900 is the final Tzade, while the 30 is the Lamed. Final

Tzade is the end of the judgment, that is, of the dream of duality; and Lamed is that this will be through the learning process. The letters Tzade and Lamed form the word Tzel: shadow. This is because it is the time of the shadow, which is the opposite of the reflection of light, since light has no shadow. Adam did not reach 1,000 years because he lacked 70, that is, 7 complete cycles necessary for his perfection.

The majority believe that it was the Perfect One - whom they encompass in the abstract idea of "God" - who took Jevah from Adam, since they have not understood the mystery or the metaphor, nor its interpretation. The rib (in Hebrew 'Tzelá') came from the 'Tzel' (shadow), which is the very reflection of the First Adam, who is also the Second Adam. For this reason he himself had to come through the 'Tzeleb' (cross) to unite in him what had been separated from him in the beginning (Tzel-B is the shadow becoming corporal, taking on the 4 elements). The Tzel of God has no darkness, since the light has no shadow, it is therefore only a reflection of itself, as the man is a reflection of the woman and the woman of the man: this is the "image", which precedes the " "likeness", and image that emanates from likeness. The Beit letter of 'Tzeleb' is the home to which we return through Adam, who is also called Ben Adam, or in English, "the son of man." Yes, for Beit means "house," and that house is the eternal congregation. Furthermore, Jevah left Adam inside paradise, but he never joined him while they were in paradise, but only once outside, because if he had joined him in paradise - that is, in heaven - they would not have dead.

They died because they were separated, and so Adam returned to unite what had been separated, and make the two one. To have separated was to have distanced one part from the other, where previously they were one. The masculine and the feminine were divided, and since then they seek to become one again. Then, after speaking of the end of all this, and of the cycle of Yom Seven, he mentions the new man who was now coming to be created: "Then Iaheveh Elohim formed the man of the dust of the ground, and breathed into his nostrils the breath of life, and man was a living being. (Gen. 2:7, KJV 60) John's version says that when creating the human, "for a long time his creation did not move or stir at all," and that Wisdom, to regain her power, sent 5 angels of the eternal realms to deceive Ialdabaot so that he breathed his power into the motionless creature, ergo "He breathed his spirit into Adam." And he adds that « Thus, the power of the Mother left Ialdabaot and entered the psychic body that had been made as the One that it is from the beginning. The body moved, and became powerful. And he was enlightened." (Secret Book of John 10:7-8)

As we see, the first human that Moses spoke of was made "image and likeness" of the gods - within the projection -, and was androgynous, but the one who came in the subsequent era was no longer produced immortal, but mortal (he had the external image of an angel, but the biological and internal resemblance of a mammal): "from the dust of the earth." This one, who is understood to have already had independent sex, was deprived of immortality, but the story also seems to have a contradiction, either in the thesis or in the chronology, since much later is when a human being is extracted. rib to make its counterpart: " And of the rib that Yaheveh Elohim took from the man, he made a woman, and brought her to the man." (Gen. 2:22 - KJV 60) The Nag Hammadi texts and the Celtic records also mention this fact: the Egyptian Christian source maintains that this was the production of a human race with the capacity for awakening consciousness, where the woman who appeared was an image of the daughter of 'Wisdom', and this was an incarnation of this celestial woman, who awakened man, in a sense of awareness of himself as a glorious and divine creature, letting him know that he was an incarnated soul, and that it really came from higher realms.

That is why Yeshua said to John: «Immediately the rest of the powers became jealous. Although Adam had been born through all of them, and they had given their power to this human, Adam was nevertheless more intelligent than the creators and the first ruler. When they realized that Adam was enlightened and could think more clearly than them, and was free from evil, they took Adam and threw him into the lowest part of the entire material realm. (Secret Book of John 10:9-10) The Kolbrin relates it by saying: "Then the fog gradually cleared and the man saw another form emerging. It was that of a woman, but one such as Fanvar had never seen before, beautiful beyond his conception of beauty, with such perfection of form and grace that he was stupefied. However, the vision was not important, it was a ghost, an ethereal being. According to this story, this Adam, called Fanvar, was at the borders of the eastern garden and saw this ethereal entity constantly, as it appeared to him deliberately.

The story goes that one day the Bothas (a savage Yosling race (intelligent semi-human beings deprived of the capacity for spiritual transcendence) attacked him while he was sleeping, and in a fierce battle he managed to get rid of them, but not without receiving a slash in his neck. side that was bleeding him, "He became mild, falling into a deep sleep and while he slept something wonderful happened: The ghost came and lay down next to [him], taking the blood from his wound to herself so it froze." for her. Thus, the spiritual creature became clothed in flesh, born of clotted blood, and being divided from her side, a mortal woman arose. In her heart Fanvar was not at rest, because of her likeness, but she was gentle, ministering to him diligently, and being skilled in the ways of healing, she made him all. Therefore, when she had grown strong again he made her queen of the Gardenland, and she was called thus., even by our fathers who named her Gulah, but Fanvar called [her] Aruah, which means 'companion'. In our language she is called the

Lady of Lanevid. It is notable that the story is known by multiple cultures, and the symbols vary little, the variations even being complementary details. Of course, regardless of the literal or colloquial events, here we are addressing their metaphysical aspect.

Another story, the Gilgamesh epic, tells us: «When Anu had heard their complaints, The great Aruru they called: "You, Aruru, created man; Now create your double; With his stormy heart make him compete. Fight each other, so that Uruk may know peace!" When Aruru heard this, A double of Anu within her conceived. Aruru washed her hands, Picked up clay and threw it into the steppe. In the steppe he created the brave Enkidu, Scion of..., essence of Ninurta. Coarse hair is his whole body, He has hair on his head like a woman. The curls of her hair sprout like Nisabal.» The human, named Adapa, had resulted from the first Mu (proto-men created by Enki to replace the work of the igigu), and although he knew the celestial abode, they deceived him so as not to reach immortality ; now, Adapa was duplicated to create a being similar to him, but with whom he had to compete, Enkidu. This very hairy human was basically a caveman, but Gilgamesh wanted to wake him up from his lethargy or absence of consciousness, so he brought a woman to him so that she would make him understand. He understood that she was flesh of his flesh and bone of his bones and he identified her as similar to him, and he was never the same again, even becoming Gilgamesh's best friend.

It is possible that this story has analogy with John's story about the last human: "Thus, Adam became a mortal human being, the first to descend and be set apart. The enlightened Afterthought within Adam, however, would rejuvenate Adam's mind," adding that "The first ruler [...] cast oblivion upon Adam. [...] Thus said the first ruler through the prophet: "I will make their minds slow, so that they cannot understand or discern."» (Secret Book of John 11:20-23) The man who came from heaven incarnated in bodies that grew.

The first generation were beings from the eternal realms who, upon seeing the harshness of this state of existence, decided to retire. Then a large percentage came from Mars, and mixed with those apes that had been genetically modified by Third and Fourth Density races from planets close to our solar system. The famous primate gene or Rhesus (Rh+) comes from them. But let's not be confused by appearances, the body is just that, an avatar. The real within the illusion is the Ba, the soul. Note the curiosity that 'Anima' is two particles: 'Ani' + 'Ma', where the European form Ani is spirit, and Ma is matter (in fact, Ani in Hebrew means 'I'). The Greek form Aima derived from the Latin 'Anima', and from that came 'Alma', so Latin introduced an element (the letter 'N'), and Spanish replaced 'NI' with 'L'.

The Separation began with the possibility (the two trees) of decision, it was followed by an idea not to be Alone, but accompanied by "something else"; then there was ego (snake); and this led to the experience (fruit) of listening to him. True Forgiveness understands that there is nothing ahead. The Bible says that a thousand will fall to your right hand and ten thousand to your left, as an expression that the Mind creates quantum reality. ACIM says that "nothing real can be attacked, nothing unreal exists. That is what the peace of God consists of", however, who is my neighbor? It is a reflection of myself, because "only" there is one. If something seems to attack you, you are actually attracting it on some subconscious, conscious or unconscious level because of your beliefs, as they are configured as arguments for the ego to blame you, because it knows that deep down there is an unconscious guilt for believing that we have separated from God. What is in front of us is the "Ezer Ka.Negedó", where Ezer is from the verb Laazor (to help), and Neged has meanings such as: declare, say or denounce; inform; flow or proceed; presence, against, in front, in front; towards, in the direction of The help for the being is for its evolution, and it consists of everything that is "in front" of us, in front, to reflect-project the distortions of our inner being.

The Archetype of the Tree

As I explain in the previous trilogy of this work, there are literalities about Genesis that are even a defamation and slander towards God. Characteristics that the apostle Paul defined as "fruits of the flesh" are attributed to it. The great work of Valentino maintains: "the Father was not jealous. For what jealousy could exist between him and his members? (Ev. Of Truth, 18) It makes no sense to tell innocent humans not to drink from a tree that is right where they live, because it would be tempting them. Other than that, why would he have something dangerous in a holy place for them? Was there no more room in creation? And not only that, how could a God of Love and Perfect create something bad?

Philip comments this regarding the topic in question: "Adam owes his origin to two virgins: that is, to the Spirit and the virgin earth. That is why Christ was born of a Virgin, to repair the fall that took place at the beginning. There are two trees in the [center of] paradise: one produces [animals] and the other produces men. Adam [ate] of the tree that produced animals and became an animal himself and begat animals. That is why Adam's [children] worship [animals]. The tree [whose] fruit [Adam ate] is [the tree of knowledge]. [Because of] that [sins] multiplied. [If he had] eaten [the fruit of the other tree, that is, the] fruit of the [tree of life that] produces men, [then the gods would worship] the man. "God made [man and] man made God." (Ev. Philip 83-84) The virgin, Parthenos or Betulah is the immaculate and unsullied reality. The Spirit is the Truth – the origin, the source -, and the "virgin land" is the Adamah. The tree of animals is the dual Etz Jaiah, and the tree of immortality is the non-dual Etz Jaiim. In reality, both are trees of life, since both are paths of experience, possibility and probability depending on the decision made. That is why "life" is also translated as "animal", since each totem expresses a lifestyle, personality and habitat. Whoever participated in the animal attitude and behavior could only educate more animals and beasts in turn. He who eats from the human tree produces humans, for this is being truly made man.

Man has worshiped statues and animals, because he wants to see outside himself reflections of understanding about his own being. He sees himself reflected in what he seeks: if he feels animal, he seeks answers from animals, and if he feels mortal he seeks answers from the immortals. What accounts for this difference? To the attitude of the ego. If you believe in yourself he wants to make you feel bad, so if you humiliate yourself he encourages this attitude of inferiority in you, but if you seek to go further he makes you feel powerless, to the extent that you can only hope to admire what he gives you. makes him believe that he is untouchable. The ego wants you to be an animal, and when you think about being a man you understand it as a being inferior to the gods, not as part of God. In this way it will pretend that you are always submitted and bowed to the feeling of unworthiness. The tree, called Etz in Hebrew, is where the wood is obtained. In the Hebrew language, both tree and wood are exactly the same: Etz. Over time, Jewish culture took the form Tzeleb to refer to a cross, but because of the shape, since in essence they identified it as Etz. Consequently, in the Hebrew mentality it was the same to say cross as tree (with regard to structure, not form). However, the analogy between tree and cross is quite notable, since the wooden cross was a symbol of death, as was the experimentation of duality. The same goes for the conceptualization of the Mind as a tree and the brain as a division of '4' with the idea of good and evil and the cross.

"[God planted a] paradise; man [lived in] paradise [...]. This paradise [is the place where] I will be told: "[Man, eat of] this or don't eat [of this, according to your] whim." This is the place where I will eat everything, since the tree of knowledge is there. He caused (there) the death of Adam and gave, instead, here life to men. The law was the tree: it has the property of facilitating the knowledge of good and evil, but it neither distanced (man) from evil nor confirmed him in good, but brought death to all those who ate of it. ; for by saying: "Eat this, don't eat this", he became the beginning of death." (Ev. Philip 94) By saying that "he killed Adam" is that he put him into the world-cosmos, and by saying that "here" resulted in giving "life to men" is that it was what caused them to come to the existence the separate projections of man in the Dream.

For this reason it is called "the law", which is what the Jews mistakenly call Torah because of their tradition. That is based on the diet on animals, which became theoretical norms, since they have had a veil. Those who can see, see, but what they cannot see, they do not see, and because of this the eyes of discernment make us understand that eating something is participating in that something and being that something. Therefore, the religion incorrectly called "Jewish" tries to follow symbolic precepts in the field of the literal, thus making it neither understand the truth nor stop being in complete darkness. In essence, it only keeps you in the slavery of religion and blindness, judging symbols as rules of life that must be followed to the letter. It is ironic that Judaism precisely influenced Christianity, and today those who want to return to the roots of Judaism and Hebrew culture take up the philosophy of the veil, which is the Hebrew literalist mentality, because the ego does not want freedom but the bondage to the symbols, that is, to the idols of matter.

If we look at the Mind as a tree, we can imagine the powerful Heimdall protecting the gates of the Yggdrasil ash tree at its foot, avoiding the attacks of the dragon Nídhoggr and a multitude of other worms that try to corrode its roots. The tree archetype has its power in the fundamental fact that it is the best expression of the pattern of Mind. The powerful Mind has the level of the roots - or archetype of the sefirotic Keter - which is the spiritual complex or access bridge to infinity, whose spectral field is the violet irradiation of the photonic electromagnetic field. These 7 levels are also reflected in the symbol of the Jewish Menorah - or candelabrum with 7 arms, or in the days of the week -, and it is necessary to see from the etheric source (or 'Ein') Keter begins, or state that produces the veil between the etheric and the spiritual. This bridge is called Bifröst in Scandinavian mythology, and it was symbolized, nothing more, nothing less, than a rainbow (symbol that once these scales are reached there will never be a world-death again). Keter connects that spiritual complex through cosmic influences, which is the shuttle switch called 'Spirit Complex', from which the Consciousness will enter the state of Totality.

The next level of the Mind continues in the roots of the tree that reach the blue magnetic field (the sefirot Binah and Chochmah), which is the second of these states that can also be understood as unconscious, which is Racial Memory, and it follows the cyan or celestial field, which is the one that coincides with the sefirot Gebura and Hesed, which is the Personal Memory. Then the trunk that Heimdall watches over rises, which symbolizes the subconscious aspects, the qualities of being, where Intuition appears, and whose color is green, and which in turn is represented with the sefirot Tiferet. Hence the tree expands by its branches, which is Intellectual Reasoning, the beginning of the conscious aspects of the Mind, identified with the yellow electromagnetic field. This is the third of the conscious parts or 'Conscious Complexities', symbolized by the sefirot Hod and Netzaj. Follow the second of the Conscious Complexities, which is orange, the sefirot Yesod, the identification with the leaves of the tree, or Emotions. Finally, there are the flowers and fruits of the tree, which are the Feelings, the archetypal idea that encompasses the sefira Malchut, and whose spectrum of light is red.

A tree is nourished by the elements below, which are earth and water, which in turn take their consciousness from the elements above that also nourish the tree: fire and air. The elements fire (that is: heat-light) and air are the highest elements of consciousness of the First Level of Consciousness of the densities of this octave. From them the water and the earth are activated. First is consciousness and light, fire; then the wind, or spirit. From them comes matter, whose first state is liquid and later solid. This example of the 4 elements is also noticeable in the myth of Yggdrasil, where it is said that this ash tree is accompanied by 4 ravens (Dáinn, Dvalinn, Duneyrr and Duraprór), understanding that the 4 elements are an analogy of the 4 angles of Heaven and of the Earth, that is, the cardinals. The 4 in this sense identifies the substance and consistency of matter, as seen in the directions, which in ancient Greek formed the initials ADAM, and, as I explain in RS2, certify the meaning of man being "made of the earth." ", specifically its 4 extremes or elements. So those 4 ravens of the Scandinavian legend can be associated with the 4 sons of Horus (Imset, Hapi, Duamutef and Kebehsenuf), who respectively represent the liver, lungs, stomach and intestines, and respond to the goddesses Isis., Nephthys, Neith and Selket.

These 4 cover the sky that Horus protects and watches over, just as the upper angles are delimited by the 4 'Heh'. On the other hand, the Yggdrasil story maintains that a squirrel called Ratatösk goes up and down sending messages, identifying the emissaries that connect the worlds, whether called angels, demons or another word corresponding to the competing culture. But what is truly intriguing about the story about the ash tree of the life of the worlds is that in its crown there is an eagle that in turn has a falcon. These two are symbols of one and the same aspect of power, transcendence and higher vision. Let us consider that the iconography of Horus, the lord of the terrestrial sky, from the top of the vault, was the falcon. He was called "the elevated one," and in Gnostic references there are quotes where the Christ is identified as an eagle on the top of the tree of life. What does this mean? That the achievement of transcendence is subject to the archetype of the eagle. This totem that has been used so much since ancient times by various cultures is nothing other than the symbol of Christ's illumination. Others used the falcon as a complement, as well as the regenerated state of the eagle to talk about death and rebirth, or express how the being must seek inner change alone and return as a new being.

This is taught in the Egyptian hieroglyphs of the bird Bennu, called the Phoenix by the Greeks. In Greek mythology the phoenix emerged from an egg, and when it matured it burned within itself and turned into ashes, from which an egg re-emerged from which the phoenix was once born. This symbol of reincarnation is preceded by that of the falcon and, above it, that of the eagle. Precisely the word Phoinix (or 'Foinix') in Greek means "palm tree", which in Catholic, Egyptian and Hebrew traditions was a symbol of life after death (not the state after disincarnation, but the elevation of being to resume life), this tree becoming the most important of those that identify immortality. And if you're wondering, who is that dragon that wants to devour the roots of the tree in Norse myth? Voila! It is the ego in the Mind; the idea of separation, constantly eating away at the unconscious with states of guilt and differentiation. The real war is internal, and therefore our spine is a bodily analogy of the Tree of Life itself, and the serpent of the ego is the Kundalini. Psitis referred in chapter 8 of the Gospel of Valentinus that this is a battle in which she was persecuted and harassed by the forces of darkness, and what is that? The projections that she herself created and from which she later wanted to absolve herself.

And fell into deep sleep

This is the unconscious guilt that comes from the Separation of the Mind from the Whole of which it was a part. In fact, it is still part of it, but it is believed to be outside. We find in life situations where we want to find guilty to blame the reason for our problems, but that reason is in our unconscious as stored guilt (master Seth explains it very well in his books transmitted to Janes Roberts, called 'The Seth Material ', a series of books that have no loss and explain many things). When Pistis says in the work of Valentino, "But they hate me without reason", it is the recognition of the subconscious Mind that there is no reason why such hatred can come, and adds that "the sin that I have committed is patent before you." What sin? There is no sin, but the Mind believes that there is sin, since the sin was separating from the Oneness, from the Father. Pistis (or 'Pistis Sofia' in writing, translated as 'Faithful Wisdom') experience it as personifications of the beings of light who came before us to these worlds as our interstellar neighbors, and were attacked by draconians and orians, who did not they are nothing but the manifestation on various levels of the ego, and its personifications.

Genesis 2:21 then says about man, "ve-ifal" (and fell). The single reference of these 4 letters encompasses the entire context. It must be understood that the uses of the word Adam identify the part of the Collective Mind that is projected into the dream, while the use of 'Yaheveh Ekohim' is the aspect that is seen producing the configurations. This is because the Ego constitutes several parts of an I, and it itself is composed of several parts. The Mind starts from several levels, but first from 3: conscious, subconscious and unconscious. However, there is also a superconscious, which in the spiritual case corresponds to the Holy Spirit and his understanding of the complete script of this film that we call the World. Adam is, then, the unconscious part, while Yaheveh Elohim is the subconscious part, Christ being the conscious part. These three are one, as a total part of the Collective Mind. It is "collective" due to the uniqueness and cooperation of the whole Mind, since the Holy Spirit - always shrewdly using symbols and sounds - hid in the meaning of "collective" the specification of "collective", or shared learning period.

Then Elohim-Adam fell, the part of the projection of human forms being at the lowest part of the hierarchy of consciousness. The Dream began and the consciousnesses that entered it forgot who they were and where they came from: "Tardemah al-Adam Ve.iishen." Man fell, if we read it another way, and that Fall came as a lethargy or state of deep sleep that came upon the entire divine human race, and they slept. He became unconscious in the dream, where the perception of prolonged continuity caused him to become disconnected from the notion of time. This is the trap of Time. When time appeared, loops and cycles were created like trusses of gears of more gears that force everyone within the projection to be tied to the periods of space-time and time-space continuity. In this way, what only happened in the Mind of God in an instant, for those who are within the dream are millions of cycles of time. For the Supraconscious Mind, Time is a self-healing effect, where the periods and cycles of the continuity of Time play the role of correcting the Errant Mind through the processes of life-existence.

Then he adds the passage: "ve-ikej achat mi.tzlaativ ve.isgar basar tajatanah", which refers to him "taking" that which was "One", and "from his shadows", "and closed up last flesh". Basically what is hidden here is that "reflections" of the Totality were taken after Man entered the dream, these reflections being "shadows" of his own distortion, that is, all kinds of "projections." Until now, light only had the reflection of light, which is emanation, since light has no shadow. Once doubt appears, the shadow of the ego appears, and where there is darkness at some level, the light passes to a lesser or greater extent. An idea that is light only produces light and radiates, but an idea that is wrong produces shadow. The light inside the sphere bounces into more light, but what is not transparent does not allow light to pass through, however, there are no reflections of the fullness of light, but shadow. Shadow and reflection are the same.

By saying that "he closed flesh" he refers to the constitution of the imperishable kingdoms, that is, of the Right Mind, which was automatically healed. The flesh or body is the complete structure or organism, and that "last" that was closed was the extension that was detached. In other words, upon seeing the problem, One took the Right Mind and closed the mental organism, justifying it, that is, reiterating its justice. The other part of the Mind did not suffer alteration by having separated the "tajtonit", or last part, which was the kingdom of Pistis Sofia. The word Tzel (shadow, projection) forms Tzelem (image in the sense of appearance), adding the Hebrew letter Mem, which is a symbol of water. For this reason man is "Tzelem" of Elohim in his "Dmut" (Gen. 1:26), since it is his physical appearance (archetype 'M') taken based on his similarity 'Dmut' (in Hebrew 'Dome '). By affirming "in its likeness" it means that the appearance stems from the fact that they are almost identical, like almost the same. However, Adam is the reflection of Elohim, which in the projection identifies the physical form of Elohim, that is, Adam is God in the dream. This is more transcendent when maintaining that it is in his 'Dmut', since the same word means "his bloods" (pronounced 'Damot'). Strictly speaking, Dmut is "figure", Damim being 'bloods', so Damot is more like the empirical identification of "his blood". Consequently, certainly "we are his offspring" (Acts of the Apostles 17:28). oh, and it is worth adding the curiosity that the voice Tzel also forms the Aramaic word Tzelah (to pray).

Verse 22 states: "ve.iben yaheveh elohim et-tzelá asher-lakach min-ha.adam le.ishah ve.ibah al-ha.adam." The form Iben is used here to say that it was "made", being 'Ben', son, so it is understood that its application comes from the idea that something derives from a similar person. Take a 'Tzelá', or the Tzelá is a part taken from the contextual structure, that is, from the integrity and core of what Adam is: "min ha.adam". The 'Min' or 'Men' form of "between" or "in the middle". And that Son or derivative of the Tzelá is a projection whose purpose is to be 'Aishah'. While Tzel is a shadow or projection, the added Ain is the superior being, since Ain (the Greek Omicron, or Egyptian Udjat – eye of Horus -) is the core of the being and identity of Adam, of the immortal being, of the Elohim-Christ-Man. The eye (Ain) is the silence of the Ein, and Adam is the eye or core reflected from it in the imperishable kingdoms (which is why the eye of Horus has its reason for being), which explains that to "look" at something is to create and perpetuate that reality. The quantum idea of the observer is a scientific way of understanding it: if you don't look at it it doesn't exist, if you observe it it exists.

If you hear something, that something induces an idea in you, and leads you to believe what it has told you. The vibration creates sound, and the articulation of letters (22 aleftau) produces words (logos), from which comes the action. From there comes the doctrine of declarations so used in Hinduism, Buddhism, evangelism - declare to receive or determine changes - and the Law of Attraction. If we see something we make it real, if we hear something we make it real, if we say something we make it real. Looking is putting the focus, attention and dedication to something. To listen to something is to pay attention, and therefore to attend, and to attend is to act. That is why it says "be careful what you hear". We learn by experience, by what we see by what we hear. The apostle Paul wrote that "Pistis comes by hearing", because we reprogram ourselves by perceiving the vibration of sound that our brain interprets as ideas and information. Even when speaking it happens, when noticing the vibration of the air passing through the throat, because that vibration reprograms us. That means that "Pistis" (perception of the infinite reality called Faith) comes by hearing. Each thing we hear will shape our beliefs, our convictions and our way of interpreting reality. That is Faith.

What we look at is where we will focus, and that is why it says, "keep your eye". The eye is what we really are, the Ain as a superior being. Hence the proverb that says that the Ain is the mirror of the soul,

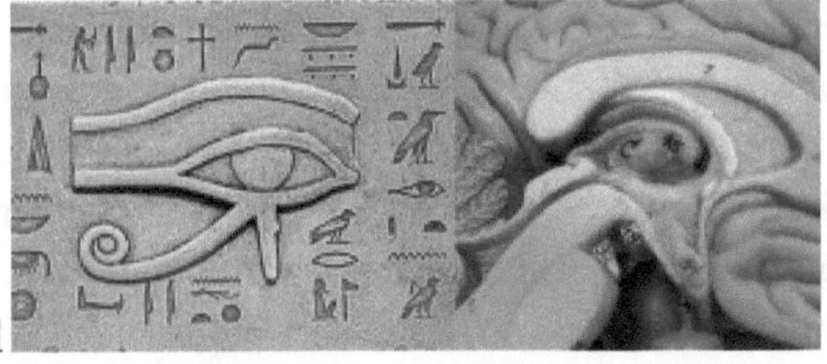

and

the one that says that if the Ain is corrupted, the whole being is corrupted.

By saying that "the lamp of the body is the eye" he refers to the illumination of the being. The being has illumination depending on the focus that the individual places on one thing or another. We have a representative eye inside the brain, which some have called the 'Third Eye', which is higher intuition and perception (the Eye of Horus). At a literal level it corresponds to the internal structure of the brain, which is the same as the parts of the Eye of Horus, or Udjat: the Thalamus, the corpus callosum, the hypothalamus, and the medulla oblongata. This connects directly with the pineal gland and the pituitary gland. These are a complete structure that represents the brain connection with the higher aspect of being. So that the letter Ain that is added to the shadow (Tzel) is later than the first derivation or the physical embodiment (Mem = Tzelem (appearance, image)). In this way, the shadow-projection (Tzel) adds the intuitive-supernatural aspect (Ain = Tzelá (rib)) producing two aspects of the same thing: Aisah (female) and Ish (male). We see because of the refraction of photons (information packets), which are the ones that make up matter. The photon is consciousness that comes from a star, which is a Logos that provides consciousness to the worlds and their consciousnesses (sub-logos).

The photon creates the atom due to the level of focused charge, and depending on its level of consciousness-energy it produces one type of atom or another, which in turn will result in the magnetism that will configure a molecule. The atom is essentially an electromagnetic structure produced by consciousness, and is more like an 'event' than a 'thing'. When you look at the atom closely it disappears, because it is only energy. When a photon is observed closely it becomes a particle, while it is not observed it produces an interference pattern, such as waves. The only plausible explanation for this is that the reality we perceive is altered-modified when we focus on it. In other words, what we do not see does not exist at a quantum level, nor does what we do not hear. What we do not focus on does not exist in these canons of our perceptible reality, and on the contrary, where we concentrate, reality is configured.

It is as if there was an instantaneous connection between the atoms/events, so as to give the appearance of there being no time or space between them. This is also valid for the Mind that observes them, since matter is before the observer. Matter and Mind constitute a unitary and indivisible whole, an unbreakable dancing whole of molten energy. These are one as are Iaheveh and Adam, a unitary concept that we can call Elohim. Iaheveh represents the elevated part of consciousness and Adam the lower, analogy of self and Neighbor. We see a similarity in the catalytic influence of Third Density which is the neighbor, outside of which are the universe of the creator and the self (that "I" has 4 sub-divisions: 1st the unmanifested Self - physical pain, the self that does not It needs a neighbor to perfect itself since it learns from the projections in its own body -, 2nd the Self in relation to the Social Self created by the Self and the Neighbor, 3rd the interaction between the Self and the devices, toys and distractions of the Self (all the inventions called neighbors, shadows, shapes or mirrors), and 4° the relationship of the Self with the attributes of the archetype of "war and rumors of war" (future projections, fears and decisions).

Sleep Levels

Before going into the matter of these two aspects of oneness that were previously united and then separated, we are going to address the understanding of the Dream and its structure of levels. Since we are in a dream, we have to understand the analogy with the other aspects that are called dreamlike, which are divided into 4 sections. In the central nervous system, 4 types of brain waves are produced: Beta, Alpha, Theta and Delta. Beta waves are 120 and 13 cycles per second, Alpha waves are 12 and 8, Theta are 7 and 6, and Delta are 5 cycles per second. The brain is a mechanism that works with waves and synaptic connections (electrical impulses of connection without contact), analogy of the Mind. Beta waves are produced during wakefulness, basically with the eyes open; Alpha waves develop during relaxation states, basically with eyes closed; Theta and Delta waves develop during Meditation and Dream processes. This is the Mind, it works in 2 phases: wakefulness and sleep. The analogy of this is with the 12 hours of the day, according to the equator. Actually, the time of the day is divided into 3 sections of 8, where sleep, as such, is usually configured in a period of 8 hours.

The brain also works in 4 fundamental states: wakefulness, REM phase, meditation and physiological sleep (see 'Transcendence', page 41). These 4 have non-conscious states and 2 muscular states. Being awake in Maia has two opposite connotations: you can be deep in sleep, which for you is being fully awake, or you can be in the dream but aware that you are dreaming and that it is a dream. The metaphysical aspect of this is that if you believe the dream, you live it, and that is why in the waking state we have muscle tone, since muscle symbolizes the perception of matter. The state of meditation in Maia symbolizes processing what is happening (questioning reality and seeking answers), or directly already searching for the disciplines to awaken Consciousness.

Sleep states, such as physiological and REM, are restorative, both for the mind and the body. Disconnections are important to encourage not being completely absorbed by mirages, which is applicable to both waking world mirages and sleeping archetypes. During sleep, consciousness works on several levels, just as it does while we are awake, except that during the day we enter it less because of the distractions of chores. Both the REM state and physiological sleep are unconscious states, since at the literal level part of our consciousness of the projection of wakefulness is disconnected and we manage to connect with the deep aspects of our being without influence from a certain part of the ego and the conscious area of separation. At a symbolic level, these states also represent two sides of the same coin, since the dream of illusion here is represented as two states of life in which one is disconnected from the reality of Truth, but in one one acts according to consciousness. and in another not. This may be due to principles of the inner voice – or Higher Self – or fear of repression, or teachings and morals instilled by parents or other educators. On the opposite level, these two states are the phases of greatest disconnection of the projection, one already absent from the distortions and the other still trying to get out of them in its search for the Truth.

Male and female

Verse 23 begins by saying, "ve.iamer ha.adam zot ha.paam etzem mi.etzmei ve.basar mi.basarei." In this verse it is the second time that the letter 'Zain' appears, relative to woman (the first was when defining Zajar ve.Nekeba, or "masculine and feminine", since man is intrinsically woman, and woman intrinsically man). It uses the form 'Zot' (Zain-Alef-Tau) three times, applicable to a woman like "she", but in the occult codes they reflect the manifestation of the mental feminine component: Zain (woman), Alef (beginning) and

Tau (end).). Zain is "weapon" in Phoenician, as it identifies our neighbor, whom the ego projected as our supposed enemy, but the Holy Spirit has created as a reflection of our diffusions of unconscious guilt. The beginning and end of Zain is the already predetermined period for the separate forms in the masculine and feminine role, the need for mirrors to see the other part of self-analysis of the Self. The passage then contains the form 'Ha.paam', which can be translated as "this time", although it also has the meanings of: disturb, drive, foot or base. He later adds the phrase, "Etzem mi.Etzmei", where Etzem (Ain-Tzade-Mem) translates 'bone', although it also means "to be powerful" or "to be strong", "to close one's eyes" or "argument".

Etzem is used in syntax to refer to something that is "the same." In the context, I would say that what was reflected in front of the figure-being was like himself, identical, that is, "his like," which became the Latin idea of "neighbor." The word Etzem comes from the word 'Etz' (tree, wood), and from Etz come forms such as Etzen (idol) or Etzeb (created object), since it is created from wood. Bone is the central structure of an organism, which gives it its consistency, and is the strongest part of the entire body. The word Etzem adds up to 200, which is the value of the letter Reish, which represents leadership, and hides the potential of the Etz (tree) and the Mem (water, matter). Then he says that it before is force of its force, structure of its structure, being of its being; It is the material projection of oneself, but seen from outside the individual self. The expression 'Mi.Etzmei' (of my bone, of my structure) adds up to 250, like the Hebrew word Ner (lamp), because through others we see our own light or our own shadows. In Greek, 250 adds up to Oinon (wine), for just as the couple gives relief and joy, so does wine in its archetypal meaning.

The sum of 'Etzem mi.Etzmei' is 450, which corresponds to the Hebrew word Neshek (kiss), a symbol of union-love. Etzem can also be seen in temura as Mi-Etz (from tree, derived from wood, coming from wood), and explains why Yeshua was crucified on a tree (cross-tree) and pierced in the "side". Justly 'sided' in Hebrew is 'Tzed' (Tzade-Dalet), with the same letter Tzade (justice, just). But apart from saying that what was placed in front of him was 'Etzmen mi.Etzmei' it was also 'Basar Mi.Basarei', where the literal translation would be "flesh of my flesh", that is, "body of my body". ". 'Basar' identifies the fleshly or physical body, written with the letters Beit-Shin-Reish, which add up to 502 (but in the reference it appears as 'Ve.Basar' ("and flesh"), so the additional sum for the particular case is 508). The form 'mi.Basarei' adds up to 552, one unit above 551, which is the sum of Tkumah (resurrection) and Taalumah (mystery). The voice Basar also creates the structure Ba-Shar, which means "in leadership", with Sar, or Shar, leader, captain or chief, whose phonetic anagram is Ba-Rash (heads, in head). The form 'Mi.Basarei' is also read in temura as 'Basarim' (meats, bodies). Another way to read this is to take 'Ve.Basar' in temura as Ba.Shor, the sound of which refers to the bull, a symbol of power in the Midwest and Mediterranean.

The next part of verse 23 refers: "le.zot ikra ishah ki ma.aish lekajah-zot", which translated is that he called her Woman because she took from Man. In notaricon we find in the initials of 6 of the final words of the verse the letters Mem, Lamed, Yud, Aleph, Kaf and Mem, which form in temura the definition 'Maalachim' (angels). Also in the penultimate 3 initials the voice 'Melech' (king), where it says "because he took from the Man". The reference 'ha.Ish' (the male) – Aleph, Shin and He – is repeated in the final letters of the phrase "he called the female because of the male", as well as the phrase "female who took from the male". The word 'Ishah' (male) adds up to 306, like Dbash (honey), which symbolizes enjoyment and longevity; the word 'Aish' (male) adds up to 311, like Shebet (scepter), a symbol of authority. The Aleph that the two words have identifies the union of Heaven and Earth; The Shin of both represents illumination, divine spark and awakening of consciousness; The 'He' of the female represents the spiritual, intuitive and expressive; the 'Yuf' of the male represents heaven, the connection with the source. Then another hidden code here comes to light: they leave the parents.

Verse 24 refers: "al-ken iaazer aish et-abiv ve.et imiv." The form 'Al-Ken' is like saying "so that", or "so", and adds that Iaazer (leave) the Aish to his father and mother. It says nothing about the female - regarding this matter - but about the male, since the Male symbolizes Adam, and the Female symbolizes Pistis. Having projected the learning system for the Mind, the configuration of several learning methods was established, among which was the system of mirrors, seeing in the other projections (forms, images) the diffusions and distortions of the Self. By separating from what was one, integral. Another separation came. The Fall consists of a system of levels, or dream layers, sub-planes of sleep. The very integration of polarities was also fractional, giving rise to polarities. There led to the complete abandonment of the Abiv (Father) and the Imav (Mother), that is, the source, the direct and fused connection with the Silence and Barbelo, the link with the Christ and Sofia the Greater, and of the being of the universe with Pistis.

This is what is meant by the appreciation that "a man leaves his father and mother." But this new experience has the objective of seeking to re-unite with its other part, which is Pistis, that is, Faith and Love. Ergo, the path of Faith and Love is the path back to completeness and absence. of the deficiencies and needs. He further adds, "ve.dabek ba.ishto ve.haiv le.basar ejad," which translates the fact that he is to "stick to his" Ishut (married, marriage) and become "one flesh." This section does not refer to the cultural concept of marriage, but to the root of the union, which at the level of the forms of the first spheres and consciousness-vibration develops through copulation. For this reason, it was written in the law of Moses that if there was no intimate sexual relationship in a marriage, it was invalidated, since Ishut – although it is translated as marriage or getting married – is the union of the forms Aish and Ishah – and therefore, the fusion of the two polarities -. I will not expand on this point since all this is discussed in detail in my work 'Naked Sex' (2017) for those who wish to delve deeper into the aforementioned matter.

What I do have to add at this point are some hidden codes, based on the fact that the sound 'Esh' means fire, and at a subliminal level it refers to sexual or passionate ardor. Ish has the Yod between the Aleph and Shin of fire, representing the elevated dividing the sexual aspect, in a sense that the light of consciousness controls the passions, but Esh is also consciousness, so that the same divine power enters into his being for it. Another aspect of Esh is heat of anger, for which consciousness controlling the being is also applied. In this way, Esh is consciousness becoming aware of itself through its passions, which it ends up controlling. Aishah has the same component introduced as Esh, but by bringing the He to the end, she structures the idea that she is "his fire", in the sense that she warms him, but also integrates his heat, be it sexual, angry or reflective. for the awareness - because what our polarizations really are bothers us from the other, and we seek in another the satisfaction that we should find in the light that is within us).

The phrase "ve.haiv le.basar ejad" has in its final letters the composition of the Hebrew definition 'Dor' (generation); The appreciation 'Le.Basar' (for meat) is made up of two phonemes: 'Leb' (heart, mind) and 'Sar' (leader, boss). The idea of Dor as a generation is because this identifies the cycle or era of the biological need for sexual union. The idea of the "heart" and the "leader" is the identification of the soul vehicle as the source of control over life. For its part, the reference 'Ba.Ishtev' has a great resemblance to the first letter of the Bible, 'Bereshit'. In Baishtev we have: BA-SH-TV; with Barashit we have: BAR-SH-IT. Both begin with 'BA' (which in ancient Egyptian was soul, but basically it is giving a home to the being), and have the 'SH' (consciousness) and 'T' (final, mark). Until that moment it is said that they were both "naked" but they were not "ashamed", but if this were about the classic myth of Adam and Eve in Eden, it is assumed that there was no one else, what did they have to be ashamed of then??

Verse 25 states: "ve.ihiv shneihem arumim ha.adam ve.ishto ve.lo itbashashu". Who were the two who were Arumim and were not ashamed? The Adam and his companion, Pistis. At the beginning of the idea of Separation and its subsequent configurations the Mind did not perceive any "problems". Arum or Eirom means cunning or nakedness, which is used colloquially to identify the sexual aspect. The form 'Itbashashu' is the conjugation of the word 'Bush', whose action in 'Busháh', with which several games of concepts can be made, among which stands out the one that 'Bush' in Aramaic will take. That is, it could be taken as "take time to realize", "slow to catch it". For example, Itbashashu could be read as a colloquialism that "he doesn't fall 6", that is, "he doesn't realize it". That would be "being in six" (Ba.Shesh), since Shesh is a strictly sexual idea in the subliminal aspects of a large part of the world's cultures: Six = Sex. To delve into this I recommend my work 'Remote Vision' (2016), where I explain many of the meanings for the type of sexual sigils and their relationship with the number 6 and its multiples, such as 666.

An additional component of this definition (Itbashashu) is that it includes within it the word 'Shabbat', as well as 'Shui' or 'Ishu', which are abbreviations for salvation; Seen another way, also with trepidation, it hides the definition 'Tashubei' (return, return). In the initials of the phrase "ve.ihiv shihem arumim ha.adam" the voice 'ha.Shua' (salvation) is hidden, but except for the first word, with the formation of the entire sentence you find the letters that form the expression "ha.ieshuav" (his salvation). The Arum form as Eirom is seen later in the characteristic of the serpent, which they define as "cunning," not "naked." What relationship can there be between nakedness and cunning? In fact they are each other's nemeses, especially when it comes to sexuality (unless it refers to the fact that the snake was also naked, being the most naked of all the creatures existing within the Dream). When the Bible says "he uncovered her nakedness" it means that he took away her virginity, and hence "cunning" as "knowing" or "knowing" is used to this day as cultural idioms in the Hebrew language to mischievously refer to the sexual act. intimate: "I knew him." Apocryphal literature maintains that the couple was "naked of any spiritual component," and the clothing that should cover their bodies is understood as a symbol of integrity, mental balance and self-respect.

The original sin

At the beginning of chapter 3 we find a direct affirmation of the existence of a species of "serpent", which is the most Arum of all the "Jait ha.Sede". Religion has never ended up with a consensus regarding whether we are talking here about a snake or a kind of angel, since if it were some type of "fallen angel" I would be comparing it to the "animals of the countryside." The text refers to the Hebrew language: "ve.ha.najash haiah arum mi.kol jaiat ha.sedeh asher ashah yaheveh elohim." The word Najash is numerically 358, the same as Mashiach (anointed, liberator), these two polarities being the same thing. The two words have notable similarities: Nachash (NJ-SH) with Mashiach (M-SH-IJ). Hebrew words are generally made up of 3 letters - or codes -, and here both have in common the Cheth (life) and the Shin (consciousness, understanding). The serpent is the archetype of the ego, and the Messiah (Mashiach) is the archetype of Savior. Since Moshiach in Greek is Christos, we understand that both words – meaning "anointed" or "anointed" – refer to the son, Christ, the image of the Right Mind.

Here we must change the word "living being" for "thoughts", since in this work we address the mental-metaphysical part of the genetic story. The "beings of the field" are thought-forms and souls of dual character, but of all of them, the most "naked" was that of the ego. That "nakedness" was the most polarized, dual and unconscious aspect of consciousness. By not putting an article here, in the word Arum, the serpent could be understood as the basis of all Jaiat ha.Sedeh, that is, the idea of the perception of separation from which all the conceptions of Death derived. Consider that the word 'Sedeh' (Sh-DH) comes from the Akkadian She-Du, referring to supernatural creatures, and was adopted into Hebrew as referring to demons. Verse 1 of this third chapter adds that the "serpent," "ve.iamer el-ha.ishah af ki-amar elohim lo taaklu mi.kol etz ha.gan," which translated would be that the serpent asks the Ishah about what Elohim would have forbidden him. The supposed dialogue begins using the word 'Af' (nose), which is a symbol of intuition and perception of life; He continues referring to Elohim and then using the word 'La' (no) and the expression 'Taaklu', which is the conjugation of the word Tojal (eat), which is a play of sounds-letters with the form 'Iejol' (to be able to, to be able to).

The snake seemed to know what the deity would have talked to the couple about, how? To eat something is to participate in it, and the ego knew that not all things that were partaken of were recommended. She was referring to "so they can't try them all?", and the woman specifies that they could eat the fruits (Pri) of the garden, but simply they should not partake of the fruit of "the one" from "in the middle" of the garden from which they were He said they shouldn't even touch it (Tiagau). This is that the mere idea of the results that this projection would give would be harmful. This happens with the Mind at all levels, because we not only create with actions but with thoughts, it is enough that an idea takes place in our mind, that we give it room. If this entered the Mind, they would die. There was masculine and feminine in the virtues of the aeonic spheres, and they were also projected below, but those above were projected opposite each other and united to create, but here, before uniting to create, they thought of creating (eating from fruit of the tree). Just as the lights-virtues emerged from the one that was total, so they were reflected before him and united to produce emanations, but Pistis listened to the ego and participated in the aeonic creation without his partner, he did not unite with him to create, but conceived of the ego

The Najash would have told the Ishah that they were not going to die: "ve.iamer ha.najash al-ha.ishah lo-mot tamot" (vers. 4) And he adds, "ki ida elohim ki ba.iom ajlajem mi. meno ve.nepkaju einijem ve.ha.ieitem ka.elohim idai tob ve.rá", which translated would be that, on the contrary, Elohim knew that at the moment they ate their eyes would be opened and they would be like Elohim knowing well and evil. In other words, you will not die, but your "eyes" will be "opened". Basically, there is no problem with making concessions regarding the Truth, since reality can be manipulated at your convenience, according to the ego. But this reasoning goes further, because it is explaining the methodology and process of persuasion of the ego in the Mind. The ego manufactures conjectures and justifications to consolidate the conviction to do or stop doing something, and for this reason we speak of "seduce". What happens then? By filling itself with reasons, the Mind agrees to participate in something, to do it, to carry it out, and it executes it, eats it.

The ego is aware that there are experiences that consciousness has not configured, so that "it does not know everything", as the ONE does. The ego's argument in this parable is that the reason for not choosing this path was a supposed jealousy of the Higher Glory over the fact that knowing all things, like the ONE, would deprive the ONE of being sovereign over all. know. The ego changes the reasoning of how he really is (jealous and envious) and attributes it to the ONE, since the ego always projects defects on the outside - blaming others - which in reality are his own defects that he wants to see in his fellow men to not recognizing their own "shame". What had not yet occurred was conception, since Adam and Javah had only seen each other as a reflection of the other - in the image of Silence and Barbeló - but they had not united. In the Higher Glories the part of oneself is seen as an integration, and when one sees oneself in front of the other they unite to produce emanations. Pistis had not yet united with Adam, who was the image of the Christ-Son, to continue producing emanations and kingdoms. When we add the words of the text " ve.napkaju enaijem" (your eyes will be opened), it gives us 450, the same as Tan (jackal) and the phrase "etzem mi.etzei" (bone of my bone), but when we look into the Calculating the order of the letters gives us 155, which corresponds to the Hebrew word Lanaah (wormwood), a symbol of bitterness; the Hebrew phrase "shikutz meshumem" (abomination of the desolate) from the book of Daniel; and "o diablolos kai satanas" (the slanderer and opponent) in Greek, from the book of Revelation.

Apart from this, 155 computes the same value as the Hebrew word 'Almah' (maiden, lady), relative to a virgin girl. This reveals to us without a doubt that the idea of "opening the eyes" was a deception and ploy of the ego to harm the man, not a revelation about a secret that UNO had with Adama. As other texts say, the Serpent (Jackal, Satan, Devil) wanted the Setite race to fall into a "desolating abomination", which was Death: an abominable (dual) and devastating state (suffering, feeling of abandonment), of bitterness. The meaning associated with the word 'Almah' is the virgin who was defiled, and that defilement was to become souls (see the phonetic association of alma with Almah). A soul is a portion of consciousness that has been individualized and that is focused within the projection of the Dream. For this reason, "ve.napkaju enaijem" (your eyes will be opened) also adds up to 144, which is the number of the redeemed (Rev. 7:4), its square root being 12: 12 x 12 = 144. 12 are the paths and their multiples, that is, the redeemed are those who have already surpassed all levels of awakening and overcoming the projection and the Ethe worlds. He had already said that the "eye" is the Setite being, and the "open" being is to disintegrate. The ego was saying something coherent, playing with words, because Adama was "opened" (retracted) like many 'Einei' (eyes), that is, like countless millions of consciousness potentials. Before they were "the eye", but now they fragmented into many "eyes", each one as an observer

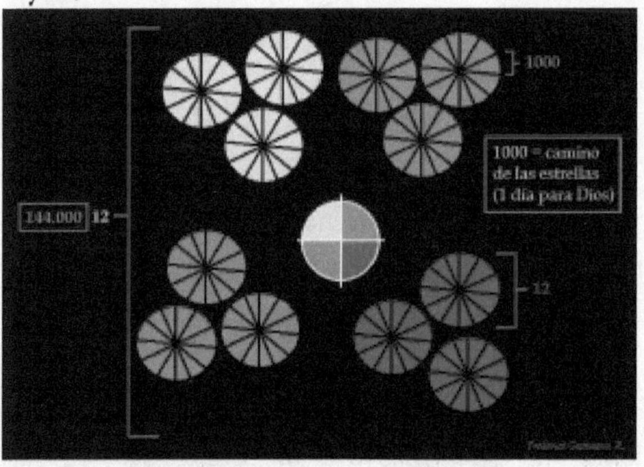

of its own reality in the cosmos.

As verse 6 says, Pistis listened to his ego and visualized that Etz as a pleasant experience (in Hebrew 'Taavah') and, furthermore, desirable (Hebrew 'Nechmad') to have wisdom, in the sense of knowing and being aware (in Hebrew 'Lehashkij'). The final letters of "ha.etz lehashkij" (tree to find out) form the word Tzel (shadow), and with the two previous letters and the three subsequent ones it forms "ed tzelem lejivan" (they will live in its image; they will live in its image), wanting to hide the code that they will experience a reality image of the superior, all things being an image of one another, and all of it an image of the superior, starting from the images of the Mind and its scale of levels. And by doing this "ve.tanen gam-la.aishah imah ve.iajal", which means that he made the "other" participate in what he tried. But here it must be emphasized that the text does not simply say 'Ish' (Alef-Yud-Shin), but Isheh, adding a He at the end of the name, and the next word defines this Isheh as the one who was with her, using the appreciation 'Ameh', which is with the letters Ain, Mem and He. It must be understood that this word refers to an 'Am' (people), not an individual. The Aisheh – or Isheh – was a collective, which was "their people." That is why the Holy Spirit prophesied about the ego: "you will not be counted among the dead, for you destroyed your people, you killed your people, but the seed of the cursed will not be counted forever." (Isaiah 14:20).

This verse is leaving hidden the fact that Pistis conceived this idea, but did not want to experience the results alone, but rather took immortal humanity with him, ergo, they were drawn into a projection that took place: the Fall. The "eyes were opened", "both eyes", as the Hebrew text of verse 7 refers, since the two represent Yin and Yang. And upon seeing this they realized that they were naked before the Truth, since they were incapable of understanding duality, the potential of Mind itself and the power that the idea of Separation would lead to. Then, the text refers, "ve.itperu alah teenah ve.iasu la.hem jageret", which translates as they made dresses with fig leaves. However, what actually happened is that bodies were manufactured. The word Taper, like "sewing", refers to manufacturing or designing, since the Mind, once the Fall took place, configured bodies of each density level for individualized consciousness according to the levels of density. When saying "leaves" it means "gods", since Allah is something light, that rises or comes from above (which is why the name Alion (the Most High) comes from). The Fig Tree is the third tree to be mentioned in the story, and it symbolizes Salvation, as it says next 've.iashu', which in Temura is read as 've.Ieshua' (and it was salvation).

What the text hiddenly refers to is that the Mind itself believed, at that moment, in the possibility of saving itself from what had just happened, placing its hope in the Fig Tree. Just as there were two previous trees - or thought-life systems -, so Salvation was configured as a new path, one of rescue, which identifies the so-called Atonement or Redemption. That path would be directed by the gods of justice (Alah), who are the first conscious part of the projection. Therefore the ancient symbol of the gods of justice was a Triangle and also a Fig Leaf within a circle. This fig leaf resembles the leaf of the vine, since the vine symbolizes filiation within the projection. The fig tree is above, and the vine below, since they are images of each other, but the vine is below - material worlds -, and therefore it crawls along the ground until deliberately the farmer (the Holy Spirit) raises it.

Such aprons (Hebrew 'Hageret') give numerically 611, as the word 'Torah'. The Torah is what gives light, instructs or teaches, which is why it is understood as a doctrine (although by Jewish tradition it is associated with the word 'Juk' (law)). In Gnostic mysticism – that is, the mystical Christianity of the first apostles – Torah was understood as the divine teachings that man receives. Judaism took the conception of the Torah as the rules of Moses, since the aforementioned were an earthly image of the superior Torah. Strictly speaking, the Torah is actually the manifestation, through commandments, of the Tree of Knowledge of Good and Evil. It is because of this that the rules imposed by Moses said "eat this" and "don't eat that", referring to participate in the distortions of the ego that configure the duality of the world-cosmos. For this reason, none of the Hebrews reached light or transcendence by obeying these laws, which Paul of Tarsus called "works of the law."

The Expulsion from Eden

The voice of conscience and the charge of guilt came to the conscience, as related in verse 8 of Genesis 3: "ve.ishmeu et-kol iaheveh elohim mi.tahalech ba.gan la.ruj ha-iom ve.itjabe ha.adam ve.ishto mi.pnei iaheveh elohim ba.toj etz ha.gan." The word that says that they "heard" fearfully forms the appreciation, 'mi.ieshuav' (of their salvation), that is, a voice within them engendered a redemptive thought. This inner voice was from the beginning until the end of time (Et) a "calling" (Kol), as Yeshua said: "many are called...", referring to "the one who hears my voice", that is, the voice of Salvation, as he also said: "I stand at the door and knock; If anyone hears my voice and opens the door, I will come in to him and will eat with him, and he with me" (Rev. 3:20). This word is also another way of referring to the Light that was in the beginning: "and there was light." The "door" that Revelation speaks of is the door of the Mind. Only the healing work of the Holy Spirit, of whom Yeshua is a part, can heal the Wrong Mind and correct it so that it is once again integrated into its rightful place, which is its natural essence, its natural state. The "supper" that the Holy Spirit develops is the interaction of Truth, which transforms the entire Mind by eliminating its judgments, distortions, idols, fears and feelings of separation.

The higher voice of consciousness, the Higher Self, called from "the other side"; The Right Mind "called the attention" of its other portion to make a "call" to integration, to return, to healing. That is what "iaheveh elohim mi.tahelech ba.gan" means, that is, the deity "who walks in the garden", which is the same as "laruach haiom", that is, in the state of fullness (Spirit). of the upper eon (the Day). Verse 8 adds that they "hid" (Hebrew 'Itjapá') from his face, that is, from the presence of consciousness (the awareness). But why? For his charge of guilt. And this happened when they infiltrated (hid) in the "trees of the garden", which are nothing other than the created worlds and their levels. This was not only manifest at the lowest levels but throughout the cosmos. Where was the man? Hidden in his reasoning. The Truth was occluded in different levels of dimension, in different levels of Mind. These cannot be penetrated as effectively as long as the state of guilt prevails, as long as forgiveness is not integrated. We must forgive each facet of the projection because in each reflection of the hologram there is a mirage that comes from that guilt, and as long as it remains unhealed (forgiven), the inner voice of redemption cannot be heard.

This is fear of the Spirit and feeling of "unworthiness" and demerit. That is why it was said that "the son of man came to seek what was lost," that is, himself. The Son of Man came into dream existence to seek what he had lost in the Fall: their lives consist of the search for Truth and atonement for their guilt. That is why Yeshua said that "the son of man has the power to forgive sins", as a basis for understanding that his sin has never been taken into account, because it is not real, it was a mere idea. The only one who is suffering from this idea is us, because we made it real, but it was real only for us. For this reason, Yeshua also stated, "your sins are forgiven," because man would only accept that God himself forgives him, since he believes that God is upset, angry or indignant because of his action. Yeshua reflects that idea of God within the projection for those who need to believe in a god who forgives them, since they are not yet able to lift their heads and feel worthy of themselves.

Hence the phrase: "the son of man came to forgive sins" - day after day, through the projections before him -, and that is, not on Yeshua, but on all of us who constitute the son of man. There is only one sin (the idea of Separation), from which they all come, and if this one is not real, no one else is. However, such is the state of guilt that the Spirit said "where are you?" (Verse 9), in the sense that "the Father seeks the Son", and with all the son feels unworthy and miserable, and wants to be hurt and afflicted for his "evil", for his sin. This is the metaphysical root of the desire for sacrifices to atone for guilt. In other words, the ego creates in the Erroneous Mind the conviction that without suffering (sacrifices) there can be no freedom, and for this reason the ideas of remission of sins (redemption) or atonement plan were born. Verse 10 is a clear example of how the Erroneous Mind enters into its lucubrations and leads to the idea of distance between Adam and Elohim. Instead of saying that he heard his voice and asked for help, what he does is that he hears the voice of the Spirit and "is afraid", because the ego established the idea of being unworthy, even when help was offered. The Mind created a hierarchy of structures and blocks to reach the Truth, and this hierarchy is symbolized in this "fear", that is, the nemesis of what in itself is Pistis (Faith) as Love.

Verse 11 relates that the Spirit asks the Erring Mind where he got the idea that he was "naked" of the Truth, and why he created based on the idea of duality, which had even warned him that the mere fact of touching it would harm him. would bring Separation and its consequences. The answers, as can be assumed, lie in outward pointing and accusations. The ego always makes the Mind look for an external culprit, instead of taking responsibility for its own actions and the consequences of its actions. The first effect is blaming others, and the next indirect effect is blaming a demon, the devil, a psychic influence, or the ego as something we cannot control. Thus man acts with his predictable reasoning: the first culprit is God because he creates the circumstances; The second culprit is the neighbor because he attacks us, provokes us and gets us into trouble; The 3rd culprit is an external agent, especially from some uncontrollable environment, be it extraterrestrial, demon, uncontrollable idea, government, education received from parents or school, chance or randomness, destiny, society, the Illuminati and the New Order World... the culprit can be abstract, literal, material, subjective, ambiguous, and even ideological (blame the pastor or priest of the church).

This is the mechanism of the ego, with which it indoctrinates ad nauseam, depriving the Mene of controlling its own thoughts and taking control of its own life. Even when the Mind seeks answers, the ego configures its dual thought system by creating new enemies, so that other nations are bad, other religions are bad, infidels and pagans are bad, sinners (unconverted) are bad, heretics are bad, gays and lesbians are bad, thieves are bad, etc. Everyone on the planet is bad, but the most ironic thing is that the one who says that is always the only one excluded from the equation, the only one who is good, and the victim of all the previous ones. Of course, no matter how close it gets to our quantum space, the closest culprit of my misfortune will be the partner, parents or children, but not oneself. The only opportunity for respite that the ego gives to look into the Mind is to foster the state of self-misery: "my fault, my fault," "I am trash," "I am the worst." This game of commiseration or victimhood is loved by the ego to continue evading its own responsibility, and it is the perfect scapegoat to escape from, to the extent that it encourages its own suffering from which the victim does not want to escape: because the party would end (the party would stop). be the center of attention, and would be forced to have to make a change, which the ego hates).

That is the incoherent ego, typical of the metaphor where they ask the inner self and the outer self, but not the ego, "why did you do it?" Because what is he supposed to respond if he's insane? He has no coherent reason for doing what he does. The ego itself is completely insane and everything it does is meaningless, nor is it worth trying to ask its reasoning or understand its attitude. Everything that derives from the ego is meaningless. Therefore it symbolizes chaos. Then Genesis 3:14 maintains that the Najash "by doing this" becomes Arur (cursed), so from then on the ego begins to identify every diabolical, satanic, demonic concept, going on to idealize every form of creature or aspect of evil.. Thus was born the myth of Satan. From this derives the mentality of scarcity and poverty, and desire for humiliation, and just as the ego is thereafter identified as the lowest idea of the Mind's system of understanding and reasoning, it is the snake at the root of the tree, at the base of the spine, at the focus of the first chakra (instinctive, reproductive-sexual and survival). Then enmity is created between the ego thought system and the intelligent reasoning system: difficulty on the path of ascent.

The ego is the source of the least spiritual thoughts of all those that reach the mind, and the one that monopolizes most of the Mind's thought system within the projection. That is what the text refers to when it says that he would be cursed among all beasts. He would always be recognized as the culprit of the state of Death. Hence the play on the words 'Gejenaj' (your belly) with 'Gehenah' (hell) in which it creeps. The ego's thought system only starts from erroneous, dead reasoning. For this reason it is also recognized that he devours (feeds) on all the consciousnesses of the cosmos (Afar = dust) as long as the illusion lasts: until the end of days (end of this eon), of time (end of Time).. Verse 12 emphasizes the enmity between those born for life and those born for suffering, which at the Mene level refers to the permanent mental conflict of the being: clash of erroneous thoughts and objective thoughts, where objective thoughts will gain ground by going to the root of the problem (the head) to correct it, while the ego will pull for the particular weaknesses (heel) of each individualized consciousness - that is, the weak points of the Mene will be what will cause suffering -.

The male and female are representing the two hemispheres of the brain in the biological mental aspect, while representing the two polarities of all things in a dualistic sense. This appreciation is applicable to all levels within the Projection. Verse 16 maintains that the Spirit signals to the Feminine Hemisphere that its desire for feminine polarity will be mitigated, and that wanting to act will be subject to the Masculine Hemisphere. In the same way, all polarity must be balanced, and depending on the straight part of a polarity (positive polarity) on the mental corrections of its counterpart, so that an aspect of duality will depend on the compensation of the opposite. Thus the hungry will depend on the satisfied, and the satisfied on the hungry; The rich will depend on the poor and the poor on the rich, the unloved will depend on the loved, and the loved on the unloved; The small will depend on the big and the big on the small. Every aspect must be placed in the balance to balance itself throughout the growth and flourishing of each portion of the individualized consciousness, just as the Collective Consciousness itself balances itself and the Collective Mind does. What should be a uniqueness will be seen unbalanced and polarized towards one of the two sides - even where something is supposed to be "positive" the dual part of itself, its shadow, must be compensated for, since both things are reflections of the same configuration created as healing scenario -.

For its part, verse 17 tells us about the "curse of the earth", that is, of the universe, since the Adamah followed Pistis in his idea of Separation. But this "curse" is not commanded by the Spirit – otherwise, it would not be spiritual – but rather a consequence given by the Mind itself. He adds that "with the sweat" of his "nose" he will work to eat, and will only see "thistles and thorns" in their state of mortal development "until the day he returns to the dust from which he came." Here is the clear understanding of the root of the frustrations and failures of Third Density life-existence. This other decompensated state is nothing but the result of adversity that you will experience as long as you do not return to oneness, as long as your Mind is not unified. The effort to obtain "food" is the search for answers and knowledge; Sweat is anguish, hard work, uncertainty, physical and psychological exhaustion and the scourge of projections of illusion. For this reason, he adds that instead of eating what he ate in Eden, he will now eat "grass" like animals, or, in other words, knowing and understanding will imply a complex work of constant struggle and patience, nourishing himself on mortal knowledge for a long time.. An example is the types of knowledge of the world, such as that of the wise Indians who understood life according to plants and trees, animals and natural phenomena. They could only go as far as understanding life under the prism of the sublunary plane, whose symbolism would be the "field grass".

When referring to the path back to the source through filiation, we must understand that Eve is the mother of the cosmos and at the same time the rescuer. That is why Adam tells her "you have given me life" (verse 20) - that is, Jevah - once he sees her standing before him. It is in the experience of "life" (Jevah), through true "faith" (Pistis) that we recognize true "Love", all of which are again and again name and attributes for the same Maia (Mary), what was a "rebellious people" (Mari-Am), or the race of "rebellion". I do not say this to pass judgment or make condemnatory approaches, because that rebellion is merely the rebellion of the ego, which causes Humanity in Glory (Adama Diamantina) to "rebel" against oneness, upon hearing its seductive voice of Separation.

Adama calls Pistis with the name of Javah, also because she would become the "mother of the living", that is, the one who would give birth not only to the cosmos, but also the one who would bring the Children of the Kingdom to this world-cosmos. She would create the life of the cosmos in its broadest and most contextual sense. Once this occurs the Spirit endows the Errant Mind with physical bodies, saying that they receive "katnot eor ve.ilbashem". Kanot are tunics, whose symbolic representation refers to the covering of the soul, that is, a vehicle or external component. If the world is material, souls would have to have a means of transportation to move in it. Curiously, it uses the word 'Eor' to refer to the skin, since it is a sound equivalent to the form Aor (light). The reason for this analogy is that the covering refers to the various light-energy bodies of consciousness in each vibratory state of density. First they made fig leaf aprons to represent the spiritual or soul body, but later that spiritual body received a wrapping for each dimensional level. At the level of the Mind this refers to the ideas of the Mind taking shape, materializing.

The integrity-light of being accompanying every soul is exposed, and the dilemma that the Collective Mind had seen the idea of duality, as the All understands and sees it (because He knows all hallucinations, although he does not conceive them as real). There was a serious risk in Adama and Pistis understanding the complexity of duality and the potential in their Mind in a fully conscious state. Only the Infinite Mind of the ONE knows the intricacies of all emanations and has the Balance of all Balances and the Wisdom of all Wisdoms, while the young Collective Mind was like a child who had just discovered a loaded revolver with no safety on. The determination-decision taken by Adama and Pistis could not be free to move within the Imperishable Realms because they would make duality an everlasting projection: it would never have an end. Even if the Mind immediately remembered its origin, the cosmos itself ran the risk of not existing with the distortion of time, and therefore duality would be a linear and permanent parameter, indefinitely maintaining the graceless state for all the fallen.

Thus it was that they were expelled from the Eternal Abodes, and they ended up in their own creation, the cosmos. That expulsion is defined as 'Ishlech' (commands, go), going to the states of samsara to deprive them of direct immortality, and in this way we all enter the perplexities (Hebrew 'Leabod') of our own creation. That Abod (work) refers to slavery, servitude, perdition, the abyss, work and other diffusions of the so-called Death – since all these words are thus translated in the Spanish Bible from the word ' Abad' and ' Abadon' -. The Fall then came from the greater planes to the mortal and suffering states, and there was a place to the "right" of the Garden, a state not far away from which Maia's story would begin. That site, called Kedem, symbolizes the ancient and initial place where humanity began in this universe, but although from there they could see the way back, a scale of hierarchies of ascending levels was set up that is protected and directed by elevated beings of pure light This Kerub whose sword turned (made spirals) protecting the path is nothing other than the guardians of the portals.

The Son of the Serpent

Chapter 4 begins by stating that Adam "knows" Javah and she conceives a son, but the suspicions that she was already pregnant lead us to analyze certain manuscripts: «...he said to the archangel Michael: "Say to Adam: 'do not reveal the secret that you know about [is] your son Cain, because he is a son of wrath. But do not [be] sad, because I will give you another son in his place, he will have to show (to you) everything you have to do. So don't say anything'."» (Revelation of Moses 3:2) A later text refers: «First the adultery occurred, then the murder. And (Cain) was begotten in adultery, for he was the son of the serpent. That is why he became a murderer just like his father, and killed his brother. For every mating that has occurred between dissimilars is adultery." (Gospel of Philip 1:46) in another we can read: «The archons approached [Norea] with the purpose of deceiving her. His supreme boss told him: "Your mother Eva came to us." But Norea turned to them and said: "You are the archons of darkness, you are cursed. You have not really met my mother, but the one you have known is in your living likeness. I am not of your progeny, rather I came from the upper world. The arrogant archon stirred with all his power and his face took on the appearance of a [...] black. Showing off his audacity, he addressed her in these terms: "It is necessary that you serve us as your mother Eva did, since it has been given to me [...]".» (Nag Hammadi Library Manuscript)

Then we find the following, which is complemented by those belonging to those found in Nag Hammadi: «And the sixth month of pregnancy came, and behold, Joseph returned from his construction work, and, entering his dwelling, he found her pregnant. And he struck his face, and fell to the ground on a sack, and wept bitterly, saying: "In what way shall I turn my eyes to the Lord my God? What prayer shall I address to you regarding this young lady? Because I received it pure from the priests of the temple, and I have not known how to keep it. Who has committed such a bad deed, and has sullied this virgin? Is the story of Adam repeating itself in me? Just as, at the very hour in which he was glorifying God, the serpent came and, finding Eve alone, deceived her, so it has happened to me." (Proto-Gospel of James 13:1) And the apostle John leaves this information: «The first ruler saw the young woman standing next to Adam and observed that the enlightened Afterthought of life had appeared in her. However, Ialdabaot was full of ignorance. So when everyone's Forethought realized what was happening, it sent emissaries and they stole Eve's Life. The first ruler raped Eve and fathered two children in her, a first and a second..." (Secret Book of John 13:5-7)

«...But this is not (the) truth nor Adam (the) Eve truth. For when they ate of the tree of knowledge, they trampled the Cherubim and Seraphim with the flaming sword. [...] having given birth to [...] descendants of the archons and their things of the world, [...]. And the females and the males, those who exist with [...] hidden of all kinds, and are going to renounce to the archons [...] a seed..." (Fragments of the Melkitzedek manuscript, from codex IX, NH) Finally, if we compare the Aramaic version, we can read: "And Adam knew Eve, his wife, who had become pregnant with Samael, [but] Adam strongly lusted for her for the angel, and transgressed her, and she gave birth to Cain." (Targum Pseudo-Jonathan, 15th century AD). The Aramaic translation explains the original Hebrew verse by implying that Adam desired to join Eve even though she was pregnant with Samael, and when Adam joined her there was adultery, since she had previously joined Samael, so that the son born of Eve was in adultery, Cain (son of Samael, and also his mother was united to another (Adam) while he (Cain) was in gestation (that is, another aspect of adultery)). This is the version even accepted by Jewish scholars, since they had the same opinion as the Targum for centuries, in light of the understanding of the Hebrew text.

One way to deduce the same thing simply is to understand that each name given to a person and place in Hebrew derives from a reason (the names were not given because they sounded nice). Kain means spear, a name that has nothing to do with the purpose that would be expected of the firstborn of man in the image of god, and therefore Cain does not appear in Adam's genealogy. Kain is derived from the form Kiná (zeal), which means that the name was given because his conception was given by an act that derived envy. Kain also derives from the form Kaná, which is to buy or acquire, to pay for someone. The Hebrew text states that Eve calls him that because she buys him from God, that is, she requests the child from God, as if he were not hers, like a mother who takes responsibility for another child or for a child who does not correspond to her. the seed or the role of genealogy, or the family mission. In Hebrew culture this was done when a child was illegitimate but was taken as adopted. It was common to pay for the ransom or life of another person to free them or give them the right to a family, social integration or inheritance.

This explains why the Hebrew word used for the action that the serpent (Samael) had with Eve was Hishian (seduce) not Matpatah (tempt). For this reason, the text maintains the relevance that they were "naked", but after that they saw that they were no longer naked. Nudity is used in Hebrew colloquialism to identify intimate sexual relations. For this reason he also says that they "knew" that they were naked, and "he knew", or that it was a tree of "knowledge", since all of them are cultural expressions that are used to this day in the Hebrew language to refer to the intimate sexual act, and They derive from the idea of sexual "cunning" or "experience." Precisely these verses say that they were "naked" (Arum) and that the serpent was the most cunning (Eirom) of all living things. Eirom and Arum are the same thing if read in Hebrew characters. These 3 races, or types of humanity, defined as Kain, Hebel and Set, are the representation of the adamantines in this world, that is, the Setites who arrived and defined themselves in one of these 3 aspects. Kain symbolizes the ego's decision to act and its faith in separation; Abel are just thoughts, but still dual or semi-dual, objective reasoning, but without the light from above; That is why Seth came later, which is the working of the Holy Spirit, the Right Mind. A thought kills; a thought is objective, but it does not save; another thought transcends the being and takes it back to the Source. It is precisely called Set pus, it is the image of the father of the Setite race.

The Later Era

The gods are deceived by projection and fall into the mirages of passions. Collective consciences produce massive self-destructions. Consciousness seeks to unify, but although your mind has ideas in common, your subconscious still feels guilty. The unification of knowledge, data, and experiences of the Mind collapse and disperse in its belief that through knowledge, without a spiritual component, plenitude can be reached. Man creates figures and idols in mortal love, configuring specialties in families, nations, tribes and races - instead of seeking filiation - and consequently the higher self disintegrates them into experiences of love so that they understand love along with knowledge. depth that they have safeguarded. What I'm talking about? The subsequent story tells us that the murderers (Cainites), which are the egoic thoughts, become evident when seeing what they cause in others (Habel), and begin to be defined (mark of Kain) and expelled (the land of Nod). Not long after, perversion arrives from heaven, manifesting that "fallen angels" are doing their thing and the social revolution unleashes a flood.

The Bnei Elohim (sons of the gods of Genesis 6:4) is the beginning of the unification of superior thoughts with inferior ones; the Collective Mind raised with the lowered. This mix makes it clear that those who believe they are more knowledgeable act under the arrogance of the ego, since knowing more does not indicate greater spirituality. Furthermore, intelligence, sagacity and knowledge without prudence, love or wisdom lead to tyranny against the weakest and most vulnerable, triggering injustice, suffering and pain. However, the Mind must be cleaned, and start from scratch (deluge). But if there is no search for Truth, the pattern will only be repeated, adding ideas, knowledge and affiliation with the interest of boasting, finding recognition, being self-centered and wanting to look superior, better or more special. This is expressed in the story of the construction of the famous tower of Babel. I can give many more examples, but from now on you will read and discern these meanings for yourself, since you have understood the metaphysical, mental, philosophical, metaphorical and spiritual roots of the stories.

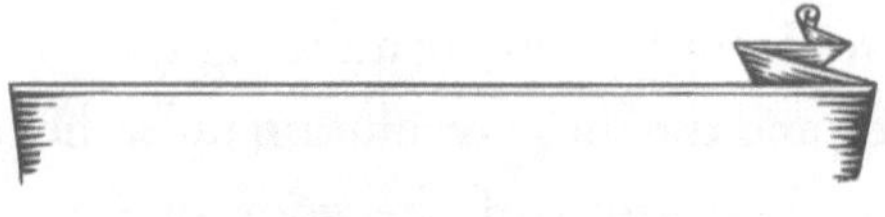

SUMMARY

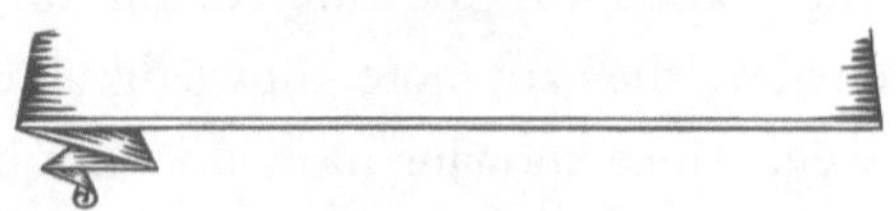

The Genesis account of Moses and current scientific theory are very much in hand with each other. They do not seem to coincide at first sight due to semantics, that is, due to the vocabulary used: the words of today and those of 3,500 years ago differ. Today we have more concise definitions for matters for which in ancient times there was no vocabulary or culture. Much less there was the incredible advance of knowledge that we have been adding decade after decade. But we must keep in mind that Hebrew and Spanish are completely different languages, and one of the dilemmas of any translator is to find not only a corresponding word, but also an idea that is similar, since many definitions do not exist in other languages. and dialects or simply mean something even opposite to what it is in another culture.

Barashit (Genesis) begins by saying in the original Hebrew: "barashit bara elohim et ha.shamaim be et ha.aretz" (verse 1). Barashit (in the beginning) bara (created, formed) Elohim et ha.Shamaim (the Shamaim) be et ha.Aretz (the Aretz). The Valentinian school (2nd century AD), which was the most similar to the teachings of Yeshua that had not yet been sullied (which happened a century and a half later when Constantine came to power) describes this fact well, defining that Shamaim as the first thing that existed, and that Aretz as the first existing reality after that initial Shamaim existed. It is important to know that this has been little known, since most worldviews start from what happened after "the light", not from what existed before it. But an "error" occurs in the chronological structuring of the text that leads to confusion if one tries to understand the story in a linear way. The following verse says that the Aretz was "disordered and empty," but the truth is that the "confusion" or "chaos" that the original Hebrew words of this verse seek to define do not evoke the cosmos, but are applicable to the cosmos in a state timeless, as "light" fills the space.

The next verse then affirms, "tié or, ve ií or" (let it be light and it was light). Scientific theory states that energy is produced by the magnitude, force, and oscillation of a wave. The energy, depending on its range, can be visible in a small fraction, like what our eyes perceive. But not all energy is visible. The vast majority is not for us, although we can perceive a certain range even if we do not necessarily see it (like heat). The Or (light) was produced after the creation of the Spiritual-Etheric Universe (the First Shamaim, or "Shamai et ha.Shamaim" (Heaven of Heavens)) and later the Aretz Universes (states of matter and semi-matter)., that appearance of light is known as the 'Big Bang'. The universes called Aretz (earth) are the Kingdom of Heaven derived from the Spiritual-Etheric Universe called Shamaim. Subsequently, in the Hebrew language, the heavens and earths (planets) that appeared later received the names, respectively, of Shamaim and Aretz. That is to say, first the ONE created a First Universe, which the Kabbalists call Ein Sof Afur, and some in classical Judaism call Arabot (wrong appreciation, since Arabot is the concept of maximum heaven within what is created in the cosmos, not before of the cosmos, but it usually serves as an idea of the upper world, or Atzilut). This First Universe is out of Time and out of Space.

The ONE Universe produced other universes, called 'Aretz', a name that our planet inherited as an image of the First Aretz. The first humanity existed there (from whom the terrestrial Adam was named) and is the origin of God, of this cosmos and of the first angels and kingdoms of heaven. Then this Aretz Universe manifested itself on a perceptible scale from which the senses were produced, called Beria, or sphere of creation. This was through an explosion of quantum mass of potential consciousness materializing in the "vacuum", and this event is called the 'Big Bang' by science. That is why it was said, "you pray" and then "you pray" (there was light). Light brought matter because Elohim became aware of himself within his own creation. This was already said by Albert Einstein in the famous formula 'E=MC2'. This means that there is a direct equivalence between mass and energy, this is the Theory of Relativity. Quantum physics verified decades later with the Double Slit Experiment that light certainly not only has mass but also functions as a wave or as a particle depending on the "observer" (who is aware of reality). This is reported by the prophet Enoch, when the Creator said to him: "If I turn my face, then all things will be destroyed." (2nd Enoch 33:5)

Light is the effect of wavelength oscillation that may be visible to us, but assuming that God made light not dependent on an observing eye, in effect the energy produced by the photon on any scale corresponds to the Hebrew word ' Or' or 'Aor', mentioned for the first time in Genesis 1 in verse 3. This means that at the time when the word Energy did not yet exist, much less Electricity (as a later manifestation of a transmutation effect of the energy), certain angels told Moses to use the word 'Or', although for the ancients it only meant "visible light." For the spiritual, Or can also be spiritual, since it is written that ONE "dwells in inaccessible OR that no eye can see," and the Messiah also said, "Let your OR shine before men so that they may see your good works and glorify them." to the Father in heaven." Yeshua uses the word Or to refer to Truth, knowledge and illumination of the being. However, if God sees everything, He sees more than the fringe of the spectrum of our carnal eyes, and sees the light of electromagnetic waves, microwaves, x-rays, infrared, gamma rays, ultraviolet rays, cosmic rays and who knows what else. photonic bands that we do not perceive.

Barashit (Genesis) tells us that the manifest Aretz (the one emanated from the first) was in "tohu va.bohu", a hendiadis used colloquially in the German language to say "madness" or "disorder". Tohu is the base particle of the Hebrew word Tehom (abyss), while Bohu is the base particle of the word Behemah (beast). The combination of both Tohu-Bohu ideas represent a single conceptual structure that means "chaos" in Hebrew. This is the antagonism of the Seder (order). Why would God create something messy? If there was nothing yet, what was it that was messed up? The first thing manifested after the universe of Elohim was an "Infinite Nothingness" (in reality it is nothing, since it is consciousness). Then the Ohr, "which was Tob (good)" acted. Tob is the adjective used by some of the apostles in early Christian culture to refer to the same person whom Yeshua called "my father", as described in the Valentinian school (2nd century AD). Tob was manifest, and for this reason he was also called Or, which other cultures translated into their own language to define the same god (which is why the word 'God' comes from the god of 'Day' (who is lord of both time and night). light)). The most notable thing about the Or is that, being the potential of the universe, it is, in turn, the entire universe. Therefore, although we try to think that light and matter are different things, in reality, matter is part of the Or, since it is energy, and not only matter but the entire cosmos. For this reason, it was discovered that the universe is sustained and connected by Dark Energy, a minor portion is Dark Matter, and the small remainder is matter as such. Although, everything is the same, because the Aether (Dark energy) and the other forms of matter are electrical charges in different degrees of wave oscillation. That is, the entire universe is Or, even where visible light does not bathe.

The narration continues, stating that the Ruach Elohim "flutters", without previously saying where that 'Ruach' (wind, spirit, force) came from. Ruach Elohim is the creative action of Tob/Or, since the photon is not just matter, as Einstein said, but a package of information, a particle of Consciousness. Consequently, the photon produced the atom by the action of an invisible "force" or "law" of motion that is sustained by the Logos because of consciousness. That Law, or Seder, moved through infinity, extending as a sphere to almost 3 x 10 at 8 meters per second. The first fractions of energy that were not yet subject to the Law of consciousness that produces the photon, were random, and consequently, "chaotic". Just like a meaningless thought when someone rambles, unlike a well-structured idea that becomes a word, a project or an action. Since the Consciousness of Elohim had not focused on certain areas, these had to wait for the time for his Consciousness to organize them.

By focusing the energy, the Ruaj "merajefet" (hovered) over the atomic mass. The first manifest effect of consciousness was the photon (Light), but the energy of Elohim focused everything on charges (positive and negative) respecting the Law, where they should not be separated from this principle. The Logos-Law is the quanto (quantum particle) that sustains the atoms. The first atomic value was called 'hydrogen' in 1783, as it means "water producer." The initial matter, as science discovered in the 20th century, from which matter in the universe is formed, and from there the masses (such as planets, comets, stars, etc.) are clouds of cosmic hydrogen gas. When compacted, it produces other reactions, leading to the formation of bodies in space. That is why it says that the Ruach hovered over the face of "ha.maim", and no one tells you why they say that it was over the water without having previously said where that water had come from.

Water in Hebrew, as a rule, is a plural word, just like Heaven. And the angels, who taught the Hebrew language, implied that the cosmos is water (which is why many cultures, like the Egyptians, called the stellar sky "sea"). Indeed, as Peter said, everything "came from water and by water subsists." For this reason the mass of celestial hydrogen is defined in the Bible as Shamaim and as Maim. The upper part that is less dense is "over there". In Hebrew "there" is 'Shama', and 'there' is 'Sham'. In space there is no above or below, and the use of above and below is really an appreciation for something light and something denser, which spiritually is the contrast between the lofty and the illusory. That is to say, the light is the cosmic gas, and the dense is the matter composing solid and liquid matter.

Then, the definition 'Maim' (waters) is formed from three abjads: 'Mem', 'Yud' and 'Mem', which means "what the sea has" (MYM = Ma YaM), or what it is composed of. the sea or what is the sea. The letter Yod goes between one Mem and another Mem (symbols of water) to remember that water exists by consciousness, not by chance, and also that there is consciousness between one material mass and another, be they planets, comets or asteroids. That is why the ancient Egyptians called the night sky "the great sea". Consequently, there were "waters" below, represented in spherical worlds with a large portion of the components of the Periodic Table of Elements expressed in various manifestations of physicality (solid, liquid and gaseous), and the same above, but in other proportions. and demonstrations. Because what is above is below. But both are also intertwined, since the planetary spheres and stars envelop and are enveloped by superior and inferior spiritual dimensions of 7 levels each.

The revolutions of the Ruach focused in space produced vortexes, and these, by Ampere's law (which is another of the laws maintained by consciousness), created toroids of consciousness and energy, attracting atomic particles to themselves and forming the planets and the planets. stars (which are the condensation, not of liquid matter, but of another form and compression of hydrogen at high temperatures in the middle of gas clouds, to fuse and create helium). The heavier elements, attracted by the focused center of consciousness, went down, solidifying, and the lighter ones were liquid (since they are maintained by energy, which is conductive), just as the even lighter ones were gaseous (which when united They project this energy in rays). Each world, depending on the level of consciousness, had more of these proportions than another. And the same with respect to the stars, whose consciousness was higher than the planets, and of the planets higher than that of the comets, and this higher than that of the asteroids. In this way the earth condensed the liquid-solid chaos within itself while the light elements were reflected in an Atmosphere. This is due to the levels of consciousness of the toroid, which creates planes in the sphere and, in them, levels of pressure.

Then, from the lower mass, it condensed one and the other progressively, separating the heavier ones from the lighter ones, giving rise to the 'Panthalasa' Ocean. And the heaviest ones later formed Pangea. But for Pangea to give rise to the continents, Shamash (Sun) brought radiation that influenced the consciousness processes of the planetary sphere, since solar radiation affects the pressures of the Earth's crust, resulting in the movement of continental plates. For this reason, when the sun appeared, life multiplied and fostered on Earth. Thus "the dry appeared," as they translate into Spanish in verse 9. And that dry (solid) part represented the Aretz of this world. The rest of the water in the planetary sphere received the collective name of 'Yamim', which is both "seas" and "days", since this sea is the image of the celestial, which was first called Or (light) and then Yom (day)., daylight).

The stars, both physical and spiritual (angels), appeared through the Ruach-Consciousness of Elohim in this cosmos to guide and be references. And the fight between good and evil began in the heavens and on the earth, as Moses said, "leiabdel ba or be ba joshej" (separate between light and darkness). For the physical is the image of the non-physical, and the material of the immaterial. Because of this, the Vedas in India called this celestial battle the 'Lilah', as in Hebrew it is 'Lilah' (night) or 'Laila'. Later, the Good One sent his angels from this cosmos to bring biological life to the Aretz worlds, like ours. They started first in the seas and then moved to land, improving the cells towards more complex structures, from the plant type to the animal type, choosing among them those that should finally stay (since they ruled out the dinosaurs because they could not live together with men. and Taninim Gdolim). They later set the stage for the biological formation of the human bodily vehicle, so that the souls of the celestial Aretz would come to this cosmos.

The purpose of these souls is to experience the universe created by the creator of the universe Aretz, called Elohim, who is us. But on that path we must understand that we separate ourselves from uniqueness in an illusory way, creating guilt, and because of it there is suffering. Then forgiveness would be the way to correct the ego's deception. They, that is, we, decided to come live in these universes and enjoy the Creation prepared for us through diverse paths of learning, but the pain entered due to the understanding that we are outside the presence of the true God and his Pleroma.. There were spiritual and semi-spiritual worlds and realms before ours – as there are in this and the other universes by untold billions – as the Midrash says, "the eternal created worlds and destroyed them." And as Rabbi Isaac ben Yakob ha-cohen said about these, there were at least 3 rebellions of angels in various celestial kingdoms before this one in which Samael (personification of the ego) opposed the ONE and became a god, saying about the Abyss, "I am God and in the midst of the Yamim I sit", and from whose school of thought Heilel ben-Shajar (Lucifer) emerged in the days of the creation of life on Earth.

The only thing that exists is Ein (which is Nothing but Everything), and from Him emanates Ein Sof (the unfathomable and illimitable), and from Him, Ein Sof Aur (limitless light). Ein Sof Aur is the reality of the Whole (Pleroma) that contemplates itself, and sees its own Mind (called Barbelo). Ein Sof Aur is the First Infinite Father, also called Silence. Barbelo is the Infinite First Mother, also called Doxa (Glory) or First Thought. Silence and Mind produced a Son (Uios) whom they called Anointed One (Christós or Mashij), and He asked for a kingdom, so they gave Him 4 stars (angels-princes), 4 assistants, 4 representatives, 3 virtues per kingdom, and multiple beings of light. That entire kingdom was called Elohim (its entirety is called 'Bosom of the Father'), and it is the Imperishable Kingdom that constitutes the first matrix universe from which 22 universes were created.

Elohim is a Collective Consciousness, composed of countless immortal lights that constitute the first race, whose first being was called Adama, that is, Man. Adama's firstborn was named Seth. Adama dwelt in the realms of the first star, called Armozel; Seth dwelt in the realms of the second star, called Oroiel; The race of Set dwelt in the realms of the third star, called Daveite. There was therefore Will and Possibility, and this allowed the Criterion, and this was deposited in the realms of the fourth star, Elelet. The three kingdoms of Elelet were Antilipsi, Eirinis and Sofias. The kingdom of Sofias was that of the virtue Pisti (which translated is 'Faith'), also called Agapi (which translated is 'Love'). The Collective Mind called Elohim, which is the same as A-Dam (divine race), thought of the idea of being something other than the Monad (the completeness of oneness) and that thought originated in Sophia. Such potential could not be contained, so it exploded towards the immensity of Ein Sof, as Moses said, "and there was light", and what science calls the

Big Bang. In this way Pisti created this universe (called Maia), but by doing so without the part of Uios (the Son), there was lack, so that the cosmos, also called world, that is, this physical universe, was produced. In other words, the idea of the ego that appeared in the Elohim Collective Mind produced a reality or dream in the Ein Sof under the principle of duality: good and evil. That is what Moses meant by saying that in Eden (the imperishable kingdoms) a tree of knowledge of good and evil was planted, and from which Adam took.

When Pisti saw what she had done, she was upset, as Moses said, "the ruach hovered over the face of the waters," and the non-dualized Right Mind referred her to help heal the dream, and therefore she was called Jevah, that is, Life, which others know collectively as the Holy Spirit. She attended to the ego (whose archetype is the serpent), and that produced the first and only sin: Separation. She could not make restitution alone, so from the Bosom of the Father - by the will of the Son - they sent 4 archangels to help her: Mijail, Gabriel, Uriel and Raphael, and then another 3. The dual idea created lights and darkness, beings of goodness. and beings of evil. Since then multiple beings from the immortal realms have come to the dream to help Pisti as part of the Holy Spirit, and others have left the darkness, repenting and consecrating themselves to the light. From the upper realms came two great lights, called Iao and Greater Melki-Tzedek, who have fought against darkness since the beginning of the world-cosmos (Maia), as Moses said, "and differentiated the light from the darkness," and wrote the apostle: "since the days of Yochanan the submerger, the kingdom of heaven suffers violence and the violent take it away" (Yochanan is the dove (Iona) that introduced us to the water of illusion but equally took us out of the illusion through of a rebirth, that is, the work of the Holy Spirit (the Iona itself).

The side of Iao and Greater Melki-Tzedek was called Yom (Day) or Iamin (Right), and that of the arch-demon Samael and his shadow Nebro was called Laila (Night) or Smol (Left). Iao's son, named Greater Adonai, recruited another god to fight the powers of Samael and Nebro that govern the fate of this part of the Milky Way from the Fifth Dimension down. That god who fought with Adonai Greater against the Left-Night was Iaheveh (known in remote antiquity as Adonai Tzabaot, or Lord of Hosts), who came to Earth more than 25,000 years ago to make the human animal an entity capable of being immortal, as Moses said, "breathed into his nostrils the breath of lives," that is, aspiration to the transcendence of being. Iaheveh returned to our planet 3,300 years ago with the aim of instructing a specific racial group with NLP codes to try to activate to another level the FOX-P2 gene that he had introduced into our DNA more than twenty thousand years ago.

Iaheveh is not a single being but a collective where more than 13 (12+1) princes have his Name. Samael sent his angels to confuse those who received Iaheveh's messages, creating dissuasive messages of elitism, nationalism, imperialism, racial supremacy, duality and religious fanaticism, which is why Samael told Abraham "sacrifice your son" - because Abraham is image in the projection of Adama in the immortal realms - but the angels of Iaheveh intervened and then made him believe that it was a test, so as not to confuse him. Iaheveh (who at the time was from the Sixth Dimension) left Earth and decades later emerged from the dream (Maia). Iaheveh was not perfect, nor were his methods perfect. Their messages were distorted because the way of thinking of humans (Third Dimensional beings) were "lower" (Isa. 55:8-9) than theirs, and were not understood.

Death appeared when Maia occurred, since it did not exist before, nor does it exist in the reality of the immortals. Samael created 7 gods, which RVA defines in Spanish as principalities or authorities; and they created 7 powers, which RVA defines in

Spanish as "powers" (where the Sirim come from), and they produced 49 arch-demons, 365 angels of darkness and many lilim, ruchot ha-temaa and shedim. Among their main hierarchy they created 36 kosmokrator to control the physical worlds and beings of lower vibration density. These beings are defined by the RVA in Spanish as "hosts of evil of this century in the celestial regions." The 7 principalities and the 7 powers mixed to create Death, which is made up of those 49 arch-demons. Many of these empires were tied in orbits and dimensional angles of the stars where they tried to control, and others were thrown into the Thalas (planes of error), and others fled as fugitives, hiding in lower planes and dimensions, many of them fleeing to planets. physical, such as the case described in Genesis 6:4 (and whose judgment is mentioned in Job 1:6 and 2:1).

However, there are no sinners. Sin is a perception of guilt from the Elohim Collective Mind that is within the Maia dream in its state of guilt. The ego of the Collective Mind seeks to create constant projections of suffering (death), from illnesses, ailments, misfortunes, "bad luck", setbacks, hell, etc. For this reason, the ego of the Collective Mind, or "the satan", materialized throughout the dream projection entities that personified its dementia, that is, the incoherent idea of believing in separating from the ONE. One of those personifications was Samael. All their forces are called Heimarmene, that is, "destiny", and their plan consists of binding the consciousnesses within their reach to Death (the state of permanent unconsciousness), and for this reason they tied their forces of darkness to the consciousnesses. For this reason, the side of Iao and Greater Melki-Tzedek have subdued and confronted them in all dimensions from the Fifth downwards, in planes and in worlds. Sin is nothing other than the connection of the being with the energy of one of the many forces of Destiny, and therefore, a bondage to

Karma (law of sowing-reaping of things called "pernicious" or "bad").

There is no one guilty of anything. They are all projections within a dream. The Mind defines, according to its particular beliefs, what the experience of being will be. When the Bible speaks of "the evil one being destroyed," it is not referring to an illusory body – because nothing real can be destroyed, and nothing that is illusory will be ratified as real by the truth – but to the wrong ideas of the Mind. dual, which the Holy Spirit will disintegrate through the awakening of consciousness. You do not understand the Scriptures as practically no one has understood them since Moses until today, because they focus on illusion (the world) and interpret it from the phenomena of a dual projection. The reality of the Scriptures is the Truth of the Mind, not of the images of a hologram: a mirage, a game of mirrors. The Mind creates the images, shapes, colors, sounds and everything that exists in the cosmos, as Yeshua expresses when he says: "it is not what goes in that contaminates, but what comes out." The bad apple is not real, the pork is not real, the corpse is not real... things are what our Mind determines them to be, so it will harm us if we believe it, or benefit us if we believe that, because everything is "according to our faith", that is, according to our scale of values and scale of beliefs.

Punishing another person is nothing other than punishing oneself, because we are ALL part of the Collective Mind. When the Big Bang occurred, the Mind created the configurations that would give rise to the spheres where the Mind itself would enter to experiment. In other words, the Collective Mind – also called Christ Mind or Universal Consciousness – set its own stage through the Logos, and retracted its consciousness throughout the cosmos. All that consciousness was focused on conscious energetic material of ether and light, producing the souls. All souls are particles of a totality that is Elohim, just as Moses said, "Adam was made afar from the Adamah." The Afar are the particles of Elohim through which Adam and Set with their children incarnated in bodies that were later created in the dimensions of this cosmos. This is the same as the quote from Moses, "he made them in his image and likeness", and that of David - and Yeshua repeated -: "I said you are elohim, and all of you are the race of Alion (of the immortals)".

For example, when the text says that "Yaheveh hardens the heart of Pharaoh", the Mind means that the Spirit (whose biblical archetype is called Iaheveh), compacts (hardens) the Mind (called "heart" in ancient Hebrew mentality) of the king (the Adama, or inhabitant of the imperishable realms). The problem with not understanding the truth is that no interpretation that one wants to make of the Scriptures will be correct; it will only be an error based on a confusion of levels and a contextual misunderstanding of facts and reality. Therefore, the Holy Spirit's interpretation of what happened in Egypt (another way of calling the dream) is that the Spirit made physical the idea of separation that the god-man (Adam-Elohim) wanted, and therefore entered the dream. and died: suffered the 10 plagues (the levels of death-separation). That dying is an archetype of the various forms and projections of suffering derived from the apparent separation. Man slowly learns this truth, or acquires the awakening of consciousness and thus enlightenment, but while he does not remember anything and is absorbed, as Moses said, "Adam entered a deep lethargy", from which at no time it was said that he had left or woken up.

For this reason the apostle Paul wrote: "if you have truly been resurrected" (at the level of the mind) "with Christ" (anointed with the truth), "set your mind" (focus of attention = reality) "on things above." " (the real), "not in those of the earth" (the unreal and illusory), because otherwise they will continue with the veil that they have always had, passing it from Jews to Christians and from Christians to neo-Jews, as happens in Western monotheism. But the veil remains the veil, until it is torn, and that tearing comes from top to bottom, first understanding the upper (the Torah of heavenly light), and then interpreting the lower (the mortal Torah, or of symbols), because it was "image of the superior", that is, ideas and symbols. We must leave religiosity and look face to face "to the living one", "to him who lives", that is, to the Holy Spirit, who "leads us to all truth", or total truth, which is the origin from which we came and to which we must return (all of this appears in the parable of the prodigal son).

Now, for the adepts of the Abrahamic religion I leave this conclusion, that the purpose of their God (Elohim) is none other than our same purpose, because the purpose is only one and the same: Return to oneness. The states of "impurity" of religion are nothing other than distortions of the Erring Mind, and you will have read in yesterday's prologue, about them it was said, "Yaheveh is a consuming fire", that is, that the Holy Spirit will take care of them. – through consciousness (whose archetype is fire) – to eliminate (consume) all erroneous thoughts from the Mind, to bring it back to the truth or state of Plenitude. That is predestination, what was determined "in Christ" already "before the foundation of the world", that is, before the Separation existed. We are part of the Mind of Christ, which is integrated into the Mind of the Whole; there is no separation, unless we make it real according to our beliefs.

Archaeological Curiosities

The medicine of yesteryear. In South America we find figures of surgical operations in remote times, both skull and heart and caesarean section. These types of interventions are not possible without specialized and advanced equipment. But the healing techniques could have been just as advanced. Very little is known about prehistoric medicine: everything we know is practically reduced to testimonies of surgical operations on bones, and these show that brain and open heart operations were carried out more than 4,000 years ago. Near Lake Sevan, in Soviet Armenia, skeletons of a people called the Jurits have been found, apparently from the year 2000 BC. A hole of about 6cm was found in one of the skulls of a woman, the result of a wound made in life. Surgeons had inserted a small plug of animal bone, and the woman survived. His own skull grew partly around the graft. Another jurit skull had a larger wound caused by a blow. Surgeons cut an area of the skull around the wound to remove the brain splinters. This patient also survived. Professor Andronik Jagharian, who studied the skulls, commented: "Considering the age of the instruments that doctors had to use, it can be said that they were technically superior to today's surgeons." Samples of cranial and rib surgery were also found in skeletons from Central Asia studied at Ashgabat University. There were clear signs that the surgical treatment had been carried out with an open heart.

Archaeologists in Iran found in December 2006 the skeleton of a woman dead some 5,000 years ago with a prosthetic eye made of tar and animal fat intact in her eye socket. Among the incredible things in medicine we have the glass lens of Eluán and the lenses for astigmatism in Assyria that are easily 6,000 years old. We assume that the first magnifying lenses are from a little over a hundred years ago, since in the mid-20th century, it was an invention that revolutionized medical science: thyroid lenses. His goal: correct astigmatism. In 1849, while the "first" lenses were being developed, archaeologist Austen Henry Layard was excavating the palace at Kalhu, the ancient capital of Assyria, better known as Nimrud. Among the innumerable pieces that he rescued, he discovered what from the beginning seemed to him to be a glass lens. Finally, the researcher put forward his conclusion after years of study: " *Everything points to the fact that it is a thyroid-shaped lens made on purpose with this shape. And lenses of this type have only one use: to correct astigmatism.* "

Dentist from 9,000 years ago. An article by Amitabh Avasthi in the National Geographic News on April 5, 2006 noted that human teeth excavated from an archaeological site in Pakistan indicated that dentistry was flourishing as recently as 9,000 years ago. Researchers excavated a Stone Age burial ground found a total of 11 teeth that had been drilled, including one that appeared to have undergone a complex procedure with a hole deep inside the tooth socket. The discovery suggests a high level of technological sophistication, although the procedure, which involved drills tipped with flint shards, could hardly have been a painful affair. " *The finding provides clear and compelling evidence that people had prior knowledge of manipulation of dental hard tissues in living people,* " said Clark Spencer Larsen, an anthropologist at The Ohio State University in Columbus, who was not part of the study. of the excavation. Scientists from the University of Poitiers in Poitiers, France, and the Guimet Museum in Paris, made the discovery. The team's findings later appeared in the journal Nature.

Caveman vanity. Once you are healthy, the next thing is to groom yourself well. The first beauty salon is more than 40,000 years old! Long before the time of Cleopatra and Helen of Troy, in the heart of Africa, the area is so mined with cosmetics to beautify men and women, that it is even recognized that nature needs a little help. The birth of beauty began at the Bomvu ridge, in the Ngwenya mountains (Kingdom of Swaziland). Here on an iron ore mountain between the Ngwenya border post and Mbabane, the capital, the first evidence of prehistoric activity was recorded in 1947.

Some 20 years later, with great excitement, it was discovered that at least 100,000 tonnes of ore had been removed prior to the start of modern open pit operations by the development of the Swaziland Iron Ore Company. And a spirit of determination gave rise to the interests of Professor Raymond Dart, of the University of the Witwatersrand (South Africa) and the Institute for the Realization of Human Potential (Philadelphia, USA). Professor Dart recommended that Peter Beaumont excavate the secrets of history in the dark. It took him about two years to do just this. Beaumont moved to Bomvo Ridge with a team of geologists and began excavating near the entrance to a small cave in Castillo Scrape. After months of hard work they were rewarded with a very important discovery: Stone tools with a carbon 14 decay date of around 400 AD, together with fragments of oxidized iron, undoubtedly establish that the Iron Age was considerable before that had been thought. More surprising was that news emerged of a site with evidence of a beauty and cosmetics center complex some 40,000 years old.

Happy Arabia. We see in the Middle East the spectacular construction that is assumed to have been commissioned by the Queen of Sheba (Bilkis). The Marib Dam is one of the great wonders of antiquity, in a territory that was populated around 1500 BC. Many of the Arab tales are compiled in the Necronomicon, although its birth among the desert cultures is still alive. Many legends and fables were not taken seriously until finds from these cities began to be unearthed. A temple called Mahram Bilquis has been found, comparable to that of Baalbeck - where there is an artifact "out of its time", since it is not possible that it was made by people of that time: The Stone of Baalbeck -, the city of Tima and many other centers. Some prehistoric places have also been found up to 75,000 years old and despite the difficulties, more than 100 other sites have been found such as Happy Arabia, which only belonged to Arab myths and legends.

Another marvel of yesteryear is the "Ed-Deir Temple" at Petra. It is believed that it was built for the Nabataean king Obodat III. The temple is 40 m high and 47 m wide; that is, it is the size of the imposing palace of a modern bank. But more overwhelming is the message bequeathed by vast literature, archaeological finds and scientific studies, regarding the "Flying Chariots of Solomon", for which he built many fueling points throughout Arabia and Mesopotamia, which became known with the name of "castles of Solomon" or "temples of Solomon" (Tajt i Suleiman), which was what Arthur Evans and Ralf Sonnenberg discovered in Crete. The stories reveal that Solomon had several "flying devices", the largest leaving and entering the Temple of Jerusalem where he rested. It had a defense system with animal figures that if someone tried to evade they were immediately mutilated. It is inconceivable that this Semitic civilization that survived thanks to the export of incense, of which the ancient world made a fantastic consumption, could move around the world without the means of transport that we have today. The fleets of King Solomon, left Eziongeber, with Phoenician crews, carrying incense to the entire world.

Don't miss out!

Visit the website below and you can sign up to receive emails whenever Frederick Guttmann publishes a new book. There's no charge and no obligation.

https://books2read.com/r/B-A-DKUGB-QCVBD

BOOKS 2 READ

Connecting independent readers to independent writers.

About the Author

Israeli writer, researcher, disseminator, documentary filmmaker and influencer. He is the writer of more than 35 books, mostly research and dissemination theses.

Read more at https://www.frederickguttmann.com.